FARMING FOR US ALL

FARMING FOR US ALL

Practical Agriculture and the Cultivation of Sustainability

TWENTIETH ANNIVERSARY EDITION

Michael Mayerfeld Bell
with
Susan Jarnagin
Gregory Peter
Donna Bauer

Photography by Helen D. Gunderson

THE PENNSYLVANIA STATE UNIVERSITY PRESS
UNIVERSITY PARK, PENNSYLVANIA

Library of Congress Cataloging-in-Publication Data

Names: Bell, Michael, 1957– author.
Title: Farming for us all : practical agriculture and the cultivation of sustainability / Michael Mayerfeld Bell.
Description: University Park, Pennsylvania : The Pennsylvania State University Press, [2024] | Includes bibliographical references and index.
Summary: "Explores the sustainability of American Agriculture, and possibilities for social, environmental, and economic change that practical, dialogic agriculture presents"—Provided by publisher.
Identifiers: LCCN 2023052088 | ISBN 9780271097480 (paperback)
Subjects: LCSH: Agriculture—Iowa. | Agriculture—Social aspects—Iowa. | Agriculture—Environmental aspects—Iowa. | Agriculture—Economic aspects—Iowa. | Sustainable agriculture—Iowa.
Classification: LCC S451.I8 B45 2024 | DDC 630.2/08609777—dc23/eng/20231212
LC record available at https://lccn.loc.gov/2023052088

First published in 2004

Printed in the United States of America
Published by The Pennsylvania State University Press, University Park, PA 16802–1003

The Pennsylvania State University Press is a member of the Association of University Presses.

It is the policy of The Pennsylvania State University Press to use acid-free paper. Publications on uncoated stock satisfy the minimum requirements of American National Standard for Information Sciences—Permanence of Paper for Printed Library Material, ANSI Z39.48–1992.

For Dick and Sharon Thompson

What you generated

Now regenerates

CONTENTS

PREFACE

"It really does look like an ocean," I thought to myself. Outside my car windows swelled waves of eastern Iowa hills, glimpsed as through portholes, rolling up and down as I sailed along Route 151. The tallgrass prairie reaches of the Midwest have often been described with marine metaphors. "Prairie settlers always saw a sea or an ocean of grass," notes Jane Smiley in *A Thousand Acres*, her novel about Iowa farming. She adds that they "could never think of any other metaphor, since most of them had lately seen the Atlantic."[1] I was tacking across this sea to Ames, Iowa, down from Madison, Wisconsin, where I now live, for the 2022 annual meeting of the Practical Farmers of Iowa. Once again, I was feeling transported into the elation of landscape I had often sensed when I lived here twenty-some years ago.

I needed to stretch my legs after a couple of hours of driving before continuing on a couple of hours more. I pulled over into the breakdown lane to savor the view from the little-used highway I had taken. I got out, the better to soak in the prairie sublime: immense, intense, engulfing every sense.

A tidal wave of smell immediately washed over me, the sour stench of feces reeking from some hog confinement facility. There was a heaviness to the air, weighing down the breath, and I staggered a bit. I scanned the horizon all around me, but I could see only fields, most resting fallow with a buzz cut of corn stubble poking out from a light dusting of snow, it being January. The source of this palpable stench was evidently miles away, and yet its waves radiated out across the landscape to where I stood, drowning the sublime. My nose whimpering, I got back in the car and sailed on.

As I drove, the smell receded (or, worse, I was getting used to it). I was soon cruising in the sublime again, captaining my ship of a Prius. But about twenty miles later, I was thrown back overboard into reality. I turned onto Highway 30 and came upon a volcanic island in full eruption—an Archer Daniels Midland ethanol plant, belching out wastewater and heat from its twisted pipes and stacks, a great gathering point of the flows of industrialism draining in from this sea, exploding from grain and gain. It, too, was immense and intense, its massive clouds reaching up and out, towering over the road but not at all sublime, at least not to me.

Iowa, is it too late for you?

The Practical Farmers of Iowa remain convinced it's not, despite the continued industrialization of agriculture that they have likewise continued to contest since their founding in 1985—and since the first edition of this book in 2004. The sustainable agriculture movement may not, as of yet, have revolutionized agriculture in Iowa, the buckle of the Corn Belt. The agricultural "middle" is still being squeezed out as the forces of industrialism tighten down even more, promoting mega-farms and micro-farms, with precious little space for what has long been quaintly (albeit not unproblematically, as I'll discuss) called "family farms." Indeed, over half of farm households in the United States actually lose money on their farm operation most years.[2] And Iowa's streams still often run brown with lost soil, and Iowa's air still often coughs with hog stink. Eaters face ever-higher prices and more uncertain supply chains, as the recent experience with COVID's effects on food availability brought more urgently to light. The current situation simply isn't working, at least not for the many. But the efforts of groups like PFI—the acronym of the Practical Farmers of Iowa—have succeeded in showing that a more agroecological way is possible, even as the buckle tightens.

Thousands of farmers are turning to PFI in response, embracing eco-friendly approaches to farming, and often in very comprehensive ways, not just around the edges of their farms and routines. In the early 2000s, PFI had some 700 members, half of them farmers—already an encouraging total in a state with, at the time, about 90,000 farms. Now Iowa is down to about 85,000 farms, but PFI is up to around 5,000 members, 90 percent of whom farm. Iowa has gone from around half a percent of farm households who are members of PFI to over 5 percent. In other words, membership has grown more than sevenfold, even in a landscape where farm numbers are shrinking.

This growing harvest of change demonstrates the basic argument of this book: that economic need is not the main barrier to sustainable agriculture. *Farming for Us All* argues that social identity, and the knowledge relations associated with identity, poses the biggest barrier. Not economics. The main crop that a farmer grows is the self, a social self, seeking acknowledgment, recognition, dignity, worth, and purpose from the relational ecology of living life, as we must, with others. To change how you farm is to change that ecology too. To grow differently is to be differently. Different friends. Different knowledge. Different tissues of trust that connect what you know with whom you know. The decisive change stems from the social relations of the heart, not the economic relations of the wallet.

Cultivation of a farm entails what this book calls the *cultivation of knowledge*: the way we nurture self and knowledge within networks of social trust. Not only farmers cultivate knowledge. We all do. We have to. Most of what we know we learn from others—yes, often tested against our own experience, but generally quite incompletely. Each of us is only one person with one life. We don't have the time or the means to first test most of what we try to put into action, or to first test what we decide is ill-advised and best avoided. We don't have time for surprise, or for failure. So we ask others—others whom we trust—about their experiences and what they have heard about the experiences of others. We trust what they know because we identify with them, as they identify with us. We want to know from them not just *what* knowledge but *whose* knowledge.

The cultivation of knowledge has a subtle yet powerful corollary: the *cultivation of the ignorable*. We ignore far more than we pay attention to. Again, we have to, living just one life with one set of eyes and ears. If you sleep 8 hours a day, you have 5,844 waking hours a year, or 350,630 waking minutes. If you spent one waking minute with every one of the 335 million people currently in the United States (and assuming that the population doesn't increase), it would take you almost a thousand years—955 years, to be exact. As of 2022, there were 240 million Facebook accounts in the United States. It would take you 684 years to spend a minute with each of those people. That is a lot of experience about the world that you are never going to be able to have even a minute to learn about. You must ignore most of that potential knowledge. And you must do so without even having a look or a listen to find out whether that knowledge would be relevant and trustable for your own life and circumstances.

We handle this problem socially, as we do most things. We look to the communities and networks with which we identify in order to, in effect, prescreen what to ignore. "Confirmation bias" is a common term that tries to get at this phenomenon. I prefer to think of it as *identity bias*, where we focus on knowledge from social networks we identify with and ignore other knowledge, whether that knowledge confirms or contests what we knew previously.[3] What we try to maintain through this form of bias is not a preexisting idea but a preexisting identity—a preexisting node in the web of social relations. Our ideas change more easily than our identities. Nonetheless, if we encounter an idea from outside our cultivation of knowledge and the ignorable, we are unlikely to accept it. Identity bias in a polarized society leads to polarized knowledge. Rural versus urban. Republican versus Democrat. Farmer versus environmentalist. Rich versus poor. Local versus expert. White versus Black. The coasts versus the Midwest. All of which can lead our cultivations to bubble into misunderstanding, conflict, lack of sympathy, and even hate.

The greatest strength of PFI is its success in bursting such bubbles. Its participatory approach cultivates knowledge for an agriculture that is not just sustainable but relational from the get-go. And when you burst those bubbles, you open up creativity. People's differences in experience and situation start to cross-fertilize one another. One plus one equals not just two, and one plus one plus one equals not just three, but solutions new to everyone. Exchanging ideas creates *grounded knowledge* that links our contexts just as one patch of ground potentially connects to every other, if we take down the fences and try to root across our differences.[4]

Take the presentation by Jason Mauck and Zach Smith at PFI's 2022 annual meeting, Saturday morning at nine o'clock in the main auditorium, talking about what they called "stock-cropping." I spend a good bit of time in agricultural circles. Although I am a college professor, my appointment is in the University of Wisconsin's College of Agricultural and Life Sciences. I co-teach an introductory agroecology course with an agronomist and an entomologist. I'm involved in a bunch of grants on (choose your adjective) sustainable, regenerative, resilient, organic, ecological, climate-smart, soil-health-promoting, and alternative agriculture. But I had never heard of "stock-cropping" and the related concept of "enterprise stacking."

Jason and Zach were taking turns with the microphone at the front of the auditorium. Jason passed the mic to Zach to explain how they (along with a couple of other farmers) came up with the idea of stock-cropping. For some years, sustainable farmers have been working out how to graze chickens outdoors with "chicken tractors," an idea originally promoted by Joel Salatin, a farmer in Virginia. Instead of raising chickens in huge confinement barns with tens of thousands of birds per building, with one or two square feet of living space each, and instead of having to bring the feed to the birds and then dispose of their manure, the chicken tractor brings the birds to the feed and the manure to the crop. A common design is a wooden coop on wheels, suitable for maybe up to a hundred birds, that the farmer periodically moves to give the chickens access to new forage. It lets the chickens live a good chicken life, outdoors with shelter close by when they want it. And it keeps the fields green and growing by having the chickens do the fertilization of the plants through their manure, slowly applied exactly when the growing crop needs it—instead of periodically spreading a layer of manure by machine, often at a rate that is way more than the crops and soil can take up and hold onto, leaching pollution into the groundwater and off-gassing nitrous oxide into the sky. The crops and soil feed the chickens and the chickens feed the crops and soil, with minimal intervention by the farmer and with minimal water pollution and climate damage.

What Jason and Zach worked out, along with Sheldon Stevermer and Lance Peterson, was basically a chicken tractor for more than one species of animal. It's more of a small moveable barn than a moveable coop. Sheep, goats, chickens, and pigs all flourish together in one unit, with individual pens for each species. All animals eat differently. By moving this multispecies tractor daily across a field, one species feasts on what the other leaves behind, maximizing ecological and economic gain. (The ecological and the economic do not necessarily have to collide. Indeed, they had better not, if we are to sustain ourselves and the land.) They laughingly call it the "cluster cluck" system.

The other key feature is interweaving this mobile barn with Jason's "strip cropping," which lays down lanes of pasture grass between strips of corn, and with his "relay cropping," in which he plants a new crop before even harvesting the current one. The strips of pasture and relays of crops help keep the ground alive and covered, holding onto the soil, and also give more light and thus more growth to the strips of corn. The "stock-cropper" barn—the "cluster cluck"—then moves down those strips and relays, fertilizing them without chemicals and "stacking" the enterprises by integrating crops and animals. Instead of just corn, you also get four species of meat to sell. Instead of monoculture, you get diversity. Instead of paying out for a big machine or a big barrel of chemicals, you get something that you can build yourself and adapt to your own farm and your own circumstances. Now that's practical.

But changing farming practices isn't like changing your socks. Jason and Zach can explain stock-cropping in about ten minutes. If you're familiar with Iowa farming, it's not complex. Yet that doesn't mean you're going to take up the practice. You have to identify with it too.

Zach explained his own identity transformation. "For the last fifteen to twenty years, I'd been in what people would I guess ascribe to the big ag space," Zach said to the crowd. He'd been selling chemicals and seed as well as raising corn and soybeans on his family's farm, which became his when his father retired in 2013. "Everything you'd see in conventional agriculture, that was what I was a part of." But something didn't fit: "At forty-two years old, I was kind of in this midlife crisis." Zach had implemented some conservation practices on his farm, yet he was feeling the constant pressure to work the land harder and to get bigger, bigger, and bigger. It didn't sit right. He was bothered that he was contributing to what he calls the "funnel of consolidation" in agriculture.

> The way it's going right now, we're just going to have a few companies left and a few farmers left at the end. My goal is not to end up one of potentially

> ten farms to farm a county and be the proud owner of an X9 combine [a top-of-the-line John Deere machine]. I just don't know if the path that we're on right now, where you just have a few left doing this, is really at the end of the day going to be good for our soil, our communities, and our general way of life.[5]

Most PFI farmers have gone through a personal crisis of some sort that shocked them out of the way they had been farming. Farming is a buzz, as this book calls it. It's not just a way to cultivate yield but a way to cultivate a self, in all its spellbinding eco-sociality. It's an all-consuming experience that catches you like surf in that ocean of hills, and that you ride as best you can on your surfboard of a farm. So much to do. So little time. So little margin for error. Doing and being merge into a *phenomenology of farming*, farming as an experience, dependent upon your knowledge cultivations and what and whom you have decided you must trust—and how you therefore envision and enact your own self. It is pretty hard to switch surfboards mid-ride.

Yet Zach did. He did because he came to question not just his trust in the knowledge he had cultivated (and what he had ignored) but his trust in the social relations of that knowledge, and in whom those relations were for and against. Farming had come to seem anti-farmer. The "conventional agriculture" he was "part of" was attacking his soil, his communities, his way of life. There are moments when we all—not just farmers—come to question the interests behind knowledge and its social relations. Sometimes it's a general growing feeling. It seems to have been for Zach. Sometimes, as this book describes, a specific incident sets that questioning off. A brush with cancer. A market crash. A crop failure. A call from the bank about your loans.

For farmers, to question the real interests behind the crops in your fields is to interrogate the many-thousand-year-old history of rural exploitation. It is to ask why there are higher poverty rates in the countryside than in the city in nearly every country in the world, the United States included. Worldwide, 83 percent of poverty is rural, even though only about 43 percent of people now live in rural areas.[6] The city accumulates from the countryside—from its crops, forests, and mines, and from the generally poorly paid people who extract those riches from the ground and from what grows upon it. For what is the city but the rural piled high?

In recent years, there has been much talk in U.S. politics about rural resentment. Some respond that the city actually doles out a steady stream of welfare for farmers—call it farm-fare—in the form of massive crop subsidy payments and the extra costs of providing services to rural places, such as broadband, the road network, and health care.[7] But when a farmer comes down with cancer, or sees how those subsidies ultimately just tighten margins and are mainly scooped up by larger farms anyway, the age-old experience of expropriation from the rural by urban-based powers leads one to question who really benefits from the labor and risk of agriculture.

Such doubt can lead to a deep breach and sense of alienation from one's own self and everything one trusts—what this book calls a *phenomenological rupture.* (My students often tease me about this term, which admittedly sounds like a medical problem in your midriff. But it's an ideological problem in your braincase.) One might recover from that rupture and return to how one was doing and being before. Or one may, in that moment of questioning, tune in to a different cultivation of knowledge and the ignorable. Like that of PFI. Like Zach did, and like Jason, Sheldon, and Lance did too.

Zach, Jason, Sheldon, and Lance are all men, White, and I believe heterosexual too. Most farmers in the United States are. But that's changing fast. Moreover, this homogeneity was never as widespread as it was once considered to be. Because of how we have conventionally defined what constitutes a "farmer," many who farm were not always thought of as farmers. PFI has played an important role in this greater welcome and recognition of the diversity of those who farm. That was already evident twenty years ago and has steadily developed since then, although we still have more joyful work of welcome and recognition to embrace.

Consider that a couple of decades ago, the 2002 U.S. Census of Agriculture tabulated that 73 percent of "farm operators" were men, almost all White, and that 89 percent of "principal operators" were men. To anyone familiar with what actually went into keeping a farm going in 2002, and who was doing that work, it was plain that these figures significantly undercounted women's labor. Household work was considered separate from farm work, and women in farming households did (and do) most of the former. The work women did in the fields, with the livestock, with the machines, and with the record keeping was commonly considered "helping," not farming. Women themselves felt constrained to understate their roles. As this book recounts, many women would describe themselves as a "farm wife" and their male partner as the "farmer." Some still do.

By the 2017 census, however, the percentage of "farm producers" who were men had fallen to 64, and the percentage of men who were "principal producers" had fallen to 71.[8] Some of that change stems from the census slightly changing its measure from "operators," which included anyone "either doing the work or making day-to-day decisions," to "producers," which included anyone "involved in making decisions" on the farm—dropping the "doing the work" part and the "day-to-day" part. Many women likely found their roles excluded by the 2012 measure. But the change also reflects the aging farm population and the common fact that women statistically outlive men by around five years and typically partner with men who are older than they are. A generation of farming men is dying off. The average age of "principal operators" was 55.3 years in 2002. By 2017, the average age of "principal producers" had risen to 58.6 years—3.3 years higher, even though the average life expectancy in the United States had risen by just 0.6 years over that time period.[9] That three-year difference works out to a significant portion of the gender shift on its own, as many older women keep the farm going as principal producer after their male partner has died.

But it's not only that. Women are taking up farming as farmers, not as "farm wives," in much greater numbers—especially in sustainable agriculture. The 2017 census tabulated 798,500 female principal producers, more than triple the number of female principal operators in 2002. Walking through the hallways and attending sessions at the annual meeting of Practical Farmers of Iowa, one sees equal numbers of women and men. Perhaps more importantly, the number of men and women presenters is basically the same. At the 2023 annual meeting, 55 men and 58 women presented, and the keynote address was by a woman. There were twelve joint presentations by women and men from the same farm household as well. At this writing, 5 of the 13 members of PFI's board of directors are women. The president is Nathan Anderson, a man, and the executive director is Sally Worley, a woman. Of PFI's 40 staff members, just 6 identify as men.

It doesn't look like this everywhere in agriculture. A colleague of mine attended the 2023 Wisconsin Corn-Soy Expo and told me that "it was like walking back in time." It was men, men, and more men, especially on stage. The Wisconsin Corn Growers Association (one of the sponsors of that Expo) has only one woman on its board of directors and only two women with voting rights on its various committees. My colleague, a woman, mainly works in sustainable agriculture. It "felt like ten years ago," she said, to be one of the only women at an agricultural meeting.

It is not just the makeup of PFI and organizations like it that has shifted. So have underlying beliefs and attitudes around gender and equity. Part of this identity transformation entailed men coming to envision their masculinity in a more open way. In the interviews my colleagues Sue Jarnagin, Greg Peter, and Donna Bauer and I conducted, we found that PFI, and the sustainable agriculture movement in general, invites what we termed a more *dialogic masculinity* in place of the monologic masculinity more typical of agriculture. The powerlessness that so many in farming experience within industrial agriculture seems to encourage monologic modes of masculinity—autonomous individualism that tries to assert complete control in its limited sphere. Monologic masculinity can serve as a kind of social-psychological antidote to dependence and impotence in the face of Big Field, Big Iron, and Big Chemical agriculture, where one is subject to the whims of the market, the banks, and the seasons and reliant on recipes marketed by the implement and pesticide dealers. A dialogic masculinity, by contrast, is what I call in Chapter 8 "a masculinity that talks, a masculinity that comes down from the tractor seat." It's not a one-way, my-way, now-outta-my-way vision of manhood, that of a controlling tough guy who cannot engage the views of others or be open about one's uncertainties and mistakes. Rather, dialogic masculinity is an interactive vision of manhood that gains efficacy through mutualism.

Central to that mutualism is the recognition of women as farmers. Farm households become households of farmers. These changes have not always come easily. Women organized through groups like the Women, Food, and Agriculture Network, which was founded in 1994 by Denise O'Brien, an Iowa organic farmer, and has now widened its mission to include advocating for nonbinary farmers as well. And there were many tough conversations in farm households. But, by and large, the men listened, at least in PFI. It has made for a more *relational agriculture* that appreciates the full social and ecological diversity of agriculture and its embeddedness in gender relations.[10]

Dialogic masculinity also changes how men engage with other men. Zach and Jason talk openly about their doubts and failures amid their successes. They synergize ideas with each other as well as with their friends Sheldon and Lance. And they don't suspiciously hide their discoveries and challenges, worried that in an individualistic and competitive marketplace you can't give your neighbor an edge. Rather, they present what they know at a PFI meeting. And they share the microphone.

Looking back now, a striking flaw in the first edition of *Farming for Us All* was something no one involved in producing the book noted at the time: the cover. The design tried to resonate with the "us all" in the title by using a gallery of five portraits along the bottom edge, including two men, two women, and a young boy. Above the gallery, the main image was of two adult hands cupping rich, dark soil (with several worms in it), which they offered to a third hand, clearly that of a child, evoking sustainability's intergenerational promise. A balance of men and women, old and young. Nice. But everyone on the cover was White.

We didn't even think of it. I'm White, my collaborators on the project were White, and the editor at Penn State University Press was White too. We were blinded by that (at the time) unremarked privilege, unattuned to the narratives the cover excluded. It is heartening, though, that such a cover would not go unremarked today. In my first discussion with the editor for this edition, that was one change we immediately decided was essential. The relationality of agriculture extends beyond the welcoming of women as farmers. PFI's recent work, and that of the sustainable agriculture movement more generally, increasingly proceeds from this wider embrace of agriculture's diversity and a deeper understanding of its troubled history. Thus the cover for this edition does not present a particular social heritage of American farming.

A century ago, there were 950,000 Black farm operators in the United States, about 14 percent of a total of 6.5 million farm operators.[11] Today, there are only 35,000 Black farm producers, about 1 percent of the current U.S. total of 3.5 million farm producers. A staggering 95.4 percent are White. A few percent are of Latino and other heritages. The disparity is particularly wide in Iowa. As of the 2017 agricultural census, the entire state has only 98 Black farm producers out of a total of 145,000. Another 187 producers are Asian, 229 are Native American, and 737 are Latino. Altogether, that's less than one percent of Iowa farm producers.[12]

The Homestead Act of 1862 set the stage for these wide disparities. With that act, the federal government awarded settlers who lasted five years and made some improvements 160 acres of land—free. As of November 2022, the average price of an acre of Iowa farmland was $11,441. Those 160-acre homestead allotments were worth, in today's money, $1.8 million dollars. That's a heck of a wealth creation starter package for the immigrant groups, mainly European, fortunate enough to be welcomed into the United States in 1862 and the years following.

They often came with very little, to be sure. Millions of families suffered for centuries under serfdom and other deprivations of feudalism. As a serf, one's person was not owned, but one was legally bound to a particular village and required to work the fields belonging to the lord of that village. You could not move to another village where, say, a different lord offered better conditions. If your village was sold to another lord, your right to a living space and your work obligations went with it. You were a chattel of the land. There was little opportunity to create wealth and opportunity for your family.

But the rise of democracy was quickly withering that land bondage. France abolished serfdom in 1789. Most German states abolished it shortly afterwards, beginning with the duchy of Schleswig in 1797. Most other European states did likewise in the late eighteenth and early nineteenth centuries. Serfdom held on longer in Russia. Finally, Tsar Alexander abolished serfdom there, too, in 1861. The very next year, the Homestead Act was signed. Fantastic timing for Russian peasants.

Slavery was also in the process of being abolished. (The 1860s were an amazingly transformative decade.) President Lincoln signed the Emancipation Proclamation in 1863, the year after the Homestead Act, and some freed folk were able to make use of the Homestead Act and General Sherman's 1865 promise of "forty acres and a mule" to the formerly enslaved.[13] But it's one thing to get access to land and quite another to keep it and to get access to the markets, the loans, and the government programs required to capitalize on it. Many racialized barriers remained. And Native folks could access the Homestead Act only if they renounced their tribal affiliations. The land steadily Whitened—especially in the Midwest.

But Iowa's beautiful land could support more diverse farmers, just as it could support more diverse crops.

Recognizing this potential for greater diversity hinges in part on how one defines a "farm" and how that definition can exclude voices from the farming conversation. In the twenty years since the first edition of this book, the sustainable agriculture movement has worked hard to expand our view of what a farm is. I remember a good friend—a PFI member and a White man—who wondered if his three-acre CSA operation counted as a "farm." (CSA stands for "community-supported agriculture," a now-popular arrangement in which households sign up for weekly shares of a season's worth of vegetables from the same grower.) He contended it would have to be considered more than a "garden," despite the huge grain farms around him with hundreds to thousands of acres each. "I think you should call my place a farm," he added, since he made a substantial part of his living from it. Yet it was something he had to defend.

But today, PFI puts a significant focus on CSAs and on urban agriculture. Some White men farm differently, of course, like my friend. Those historically excluded from farming are especially likely to. There were nine presentations about CSA at the 2023 annual meeting, almost all by women. The keynote speaker, Donna Pearson McClish, gave a presentation about her experience as a Black, multigenerational urban farmer in Wichita, Kansas. In other words, broadening the understanding of what a farm is immediately broadens our understanding of who a farmer is.

The rise of "urban agriculture"—a phrase that no longer seems paradoxical—has been particularly important for welcoming and recognizing the contributions of diverse farmers. It's an enormous financial challenge to get into grain and livestock agriculture, as the margins are low and the land base required is huge.

Urban agriculture, including peri-urban agriculture, typically focuses on fresh vegetables. A farmer can make a living from them with less land. True, farmable land in and near cities is much more expensive because of development pressures. But people find little pockets here and there, often with the support of government and nonprofits. The city is increasingly becoming a place for growing, not only for eating. Black, Latino, Hmong, Jewish, Muslim, and Catholic farmers, many of them women and many of them young, have all been creating spaces to plant and to continue to cultivate their traditions of agricultural knowledge. Along the way, noncommercial agriculture has been rebounding in cities as well, as people of all heritages rediscover the joys of growing—whether in their backyards, with pots on their apartment balconies, or in the many community gardens that have been springing up in parks, schoolyards, and previously neglected city lots.

But there are other terms we still need to contest. There has long been great concern in sustainable agriculture circles about the fate of the "family farm." Much of that concern has been a way to challenge the increasing industrialization of agriculture, which shoves more and more small farms into the funnel of consolidation that Zach talked about, eroding both the soil and the economic basis of small- and mid-scale agriculture. There is conceptual danger in that term, however. It conjures up a particular vision of agriculture's social basis: that it ought to be organized around the heterosexual and the nuclear.[14] Say the phrase "family farm" and a slide show starts up in the mind, with images of a solid-looking man in a feed cap and jeans standing by the barn with his wife—she perhaps in jeans, too, but with at least a ghost of a vision of a gingham dress—and two to four kids, five to fifteen years old, the younger ones blonder, all arranged in a group by height, and likely everyone White. Not American Gothic so much as a rural Brady Bunch.

Such mental slides have some striking absences. Families, like love and farming, are diverse. We organize ourselves into many more forms of family farm than a heterosexual couple with a thousand acres and an X9 combine. Gender is diverse. Sexuality is diverse. Farming is diverse. And what makes a family doesn't necessarily have to do with any of those.

So should we just recognize that "family" comes in many varieties, like so many fields of flowers, and then, noting that, carry on with the term "family farm"? That would be better, but we would still need to trouble the phrase. Is it really "family" that we should be working to protect and enhance? Family is great, and I cherish my own. But sustainable agriculture, like the good things in all of social life, has many other forms of social organization than the family. Community-supported agriculture is one example. So too are farms organized around heritage traditions. The Afro-Indigenous farming traditions of New York's Soul Fire Farm. The Jewish farming traditions of North Carolina's Yesod Farm and Kitchen. The Muslim farming traditions of New York's Halal Pastures. The Catholic farming traditions of the Dominican Sisters of Sinsinawa, Wisconsin.[15] The many Hutterite colonies of Great Plains states and Canadian provinces. Sustainable farms may also be organized as worker-owned cooperatives, such as Iowa's Humble Hands Harvest. Or they may spring up from farm incubators, such as Wisconsin's Farley Center. And there are possibilities we have not even conceived of yet.

Plus the term "family farm" leaves out of view more than a third of those who farm. In addition to the 3.5 million farm producers in the United States, at least 2.2 million "farm workers" earn wages through farming—75 percent of them foreign born, mostly from Mexico.[16] It is curious that common language would make this distinction. All "farm producers" do at least some of the work on farms (even if it is only the work of decision-making), and all "farm workers" help them produce.[17] Farmers work and farmworkers farm. Fundamentally, the distinction rests on who has gained rights to a farm's land base, through owning it, renting it, leasing it, signing a crop-share contract, or engaging in some other means of land tenure. Such language is thus a curiosity of capitalism. And perhaps of racial and ethnic privilege as well, for although 95.4 percent of U.S. farm producers are White, only 24 percent of "farm workers" identify as White.[18]

Used in this way, the term "farmer" renders anonymous the work of millions who are already largely invisible. This erasure carries on into how farm organizations define themselves. Farm Bureau provides basically no services to—as we should more accurately term them—farm employees.[19] Their focus is almost entirely on farm employers, those they term "farmers." The National Farmers Union in the United States, alas, does little better, stating that its mission is "to protect and enhance the economic well-being and quality of life for family farmers, fishers, ranchers and rural communities."[20] Its "Fairness for Farmers" campaign contests how "decades of consolidation in the agriculture industry have devastated family agriculture"—a worthy goal in my view (and I am a Farmers Union member), but one that does not also encompass the concerns of farm employees, many of them migrants with uncertain legal status and abusive working and living conditions.[21]

Some farm organizations have been working to integrate the concerns of farm employers and farm employees (albeit usually retaining the language of "farmers" versus "farm workers"). The Rural Coalition is one. Via Campesina is another. The National Farmers Union of Canada explicitly provides a membership category for farm workers. In the early twentieth century, the Farmer-Labor

Party of the United States went even further and worked to unify farmers with all those who labor—rural and urban.[22] Imagine if the major political parties sought to build common cause with farmers and labor. Much-needed unity would result both within rural areas and between the rural and the urban.

Closely associated with the term "family farm" is another term we should trouble: "century farm." Several states operate programs that award certificates or other forms of recognition to farms with, as the program in Iowa describes it, "consecutive ownership within the same family for 100 years or more of at least 40 acres of the original holding."[23] Families that maintain ownership for 150 years or more can be enrolled in Iowa's "heritage farm" program, or what Wisconsin calls a "sesquicentennial farm." Families take great pride in these designations, and given all the pressures on farming, both social and economic, such continuity is remarkable. But these programs do not ask some potentially painful questions. Whose land was it before these farms were developed? How did these families gain their ownership, and what privileges of life supported them in retaining it?

Because there were farmers in Iowa, Wisconsin, and elsewhere in the United States well before these century and sesquicentennial farms were established. First Nations folk had already been farming here for millennia. A couple of hundred are still farming in Iowa, as I noted in the previous section. Many more would if they had more of their land and wealth restored. The main tribes in Iowa were the Ioway, for whom the state is named, and the Sioux, for whom Sioux City, Iowa's fourth-largest city, is named. Some fifteen other tribes also lived in Iowa, including the Sauk, the Meskwaki, the Ho-Chunk, and the Potawatomi. Few remain today—only about fifteen thousand tribal members, one of the lowest totals of any state.[24] There is just one small reservation entirely or mainly within Iowa, the Meskwaki Settlement.[25] Iowa's fertile farmland was too tempting for settlers and the agricultural industry. The legacy of coercion and violence is heavy indeed.

Perhaps we also need to speak of—and celebrate and restore—"millennium farms." Even "millennia farms."

There's an old, sardonic joke that farmers sometimes tell. "Want to make a small fortune in farming? Start with a large one." This kind of grim humor arises from the sense of crisis that most in farming feel, farm employers and farm employees alike. Even for those advantaged by the Homestead Act of 1862 and other privileges, like loans and market access, farming is rarely an easy livelihood. Yes, identity and knowledge mutually cultivate the successes of the sustainable agriculture movement. Nonetheless, huge economic challenges remain.

It's no coincidence that Practical Farmers of Iowa began in 1985, in the midst of the infamous "farm crisis" of the mid-1980s, when farmland values crashed following a temporary embargo on grain exports to Russia. The farm crisis led more people to question the wisdom of the industrial model of agriculture and its unending appetite for land base, due to the lower margins that come with industrialism's tendency toward overproduction. Not only was production going up but farm size was, too, because if your margin goes down, you have to make it up by producing more. The average size of an Iowa farm in 1950 was 169 acres, little more than the original 140-acre allotments of the Homestead Act. By 1982, it was 283 acres, a 67 percent increase.[26] The number of farms in Iowa collapsed accordingly—from 203,159 in 1950 to 115,413 in 1982. Farmers were feeling threatened. They still are. As of 2020, the average Iowa farm is 360 acres, and there are 85,000 farms.[27] In reality, the farm crisis never ended. And it began well before the 1980s.

So what's a farmer to do except to follow U.S. Secretary of Agriculture Earl Butz's often repeated advice from the 1970s and "get big or get out"? Because as a later U.S. Secretary of Agriculture, Sonny Purdue, put it in 2019, "In America, the big get bigger and the small go out." Appalling as these statements were, especially from secretaries of agriculture, they referenced a powerful force farmers must nonetheless contend with: the dynamics of that funnel of consolidation.

More land isn't the only way to increase your production of low-margin farm commodities, though. You can also do it by investing in machines, chemicals, and seeds that pump up the volume. And you might get a temporary edge—until your neighbors do it too. Then everyone is stuck with higher costs and higher production, and thus lower margins, as prices fall in response to the higher production. Desperate, you try the next new bit of techno-wizardry, and the cycle repeats as your neighbors try it too. The agricultural economist Willard Cochrane long ago called it the "treadmill of technology."[28] Once you're on it, it's hard to get off—unless you're simply forced off, your profit squeezed to nothing. Then your land gets bought up by others, and the fewer, bigger farms repeat the cycle all over again. Perhaps in the end there will be only one farm left, a huge conglomerate that controls everything.

There's also another treadmill you'll have to find your balance on: the subsidy treadmill. In recent years, the U.S. federal government has been paying out around $20 billion a year in various subsidies for different crops and the different complications and issues that farms face—and as much as $45 billion in 2020.[29] Using the $20 billion figure works out to about $10,000 a year per farm. That's a lot of farm-fare for you and your fellow farmers.[30] About $4 billion of those subsidies are for conservation programs, largely for promoting sustainable practices. That aspect of farm payments is welcome. But in terms of penciling out a living from farming, if other farms are getting this support, too, it enables all of you—or, better put, compels all of you—to cut profit margins that much further. Because if one of you does, the rest eventually will have to as well.

You can try to maintain your footing on these two treadmills in some other ways. You might cut the wages of your farm employees. You might sacrifice your stewardship of the land, water, and climate and forgo the conservation subsidies for the crop subsidies. You might do whatever you can do—and perhaps some things you really shouldn't and don't want to do—in the midst of this permanent farm crisis, trying to "stay in the game" as long as you can. But of one result you can pretty much be sure: overall production will go up in the process, even if few farmers are making any more money. Thus in many ways it is more apt to call the overall situation a "treadmill of production."[31]

Now consider the specific regional context of your farm. Factor that in and you're looking at what *Farming for Us All* calls the *farmer's problem*. If you had a good year and got good yields, likely other farms like yours in your area did as well. They were on the treadmill of production too. Consequently, you may have plenty to sell, but prices are bad. So you don't make much. If you had a bad year with bad yield, likely other farms like yours in your area also did poorly. They probably made much the same bet with technology—and with the banks, so that they could afford it. They probably also filled out the same paperwork and went after the same farm subsidies. So prices may be good because production was low, but you don't have much to sell or to claim subsidies for. You still don't make much. You really only do well when you have a good year and most other farmers like you don't. And that doesn't happen very often.

Consequently, as I mentioned earlier, median farm household income—the midpoint farm in the range of farm household incomes—is typically negative. 2021 was a relatively good year, and the median farm household netted just $210 from farming.[32] That's not a typo. In 2020, the median was a $1,198 loss. Most U.S. farm households aren't starving. Median farm household income from all sources is around $85,000 to $90,000, depending on the year. But pretty much all of that income is from off-farm employment. Driving trucks. Stocking shelves at Walmart. Serving as a nurse at the local health clinic. Teaching at the local community college.

There is yet one more challenge. Except over land base, you're really not in competition with the farm next door. The value of your production, and what you get back for it, has little to do with your neighbor. That value mostly reflects how far removed you are from eaters. It mostly has to do with what Katharine Legun and I have suggested calling *conducers*: those who conduct in the market.[33]

In between producers and consumers stretches a vast and bulging middle of processors, wholesalers, and retailers that serve as the conduits between producer and consumer—a position of power that also enables them to be conductors who organize and orchestrate the conduits, mainly with their own interests in mind. Every year the USDA calculates the share of the food dollar that actually gets back to farmers. Every year it goes down a bit further. At this writing, based on figures from 2021, the farm share is the lowest ever: 14.5 cents.[34]

Which is part of why there are any farms left at all. Investors and conglomerates are snapping up more and more land, which jacks up farmland prices, speeds the treadmill of production even more, and continues the steady decline in the number of farms. They are discovering how to use farm subsidies to become subsidy farmers. But historically, conducers have been cautious about vertically integrating right down to the level of the producer. Conducers would rather not have to take the risks of the farmer's problem. And they want leverage over those who are willing to take those risks. Conducers want their large fortunes to get even larger, not smaller. As the lyrics of a century-old Farmers Union song put it, "the merchant is the one who gets it all."[35]

But what about feeding the world? Isn't the treadmill of production—greased by the marvelous efficiencies of modern conduction, able to span the globe and ship oranges from South Africa to New York and corn from Nebraska to Cape Town—a good thing overall? Yes, some farmers go by the wayside, but it's crucial that we increase our food production, right? "One U.S. farm feeds 166 people annually in the U.S. and abroad," says the American Farm Bureau. "The global population is expected to increase by 2.2 billion by 2050, which means the world's farmers will have to grow about 70% more food than what is now produced."[36] Isn't that true?

No. We don't want to feed the world. We want a world that is fed. We want a world free of the scourge of hunger, yes, but U.S. farms don't feed the hungry. They actually feed the well-fed and the overfed. Because U.S. farms don't give their production away. They sell it. Less than 1 percent of U.S. farm exports goes to the hungriest countries in the world.[37] That's because these are all very poor countries. Haiti. Yemen. Ethiopia. Afghanistan. Namibia. The main cause of hunger is poverty, and the poor don't have money to buy food from U.S. farms. Because they are poor.

The United States does provide some free food assistance in times of extreme need and for those facing famine. We are the world's largest provider of international food assistance—about a third of all food aid. In recent years, we've been providing about $4 billion a year in food.[38] But the total value of U.S. farm exports currently runs at about $200 billion a year.[39] In other words, we provide about 2 percent of our farm exports as food assistance, and we also sell about another 1 percent to very poor and hungry countries. Moreover, we only export about a fifth of our total agricultural production to begin with.[40] Most of what we grow we consume within the United States. That 2 percent we provide and 1 percent we sell to the hungry works out to 0.6 percent of all that we grow. That does not amount to feeding the world.

Besides, if you don't have much money, your food supply is far more secure if you can provide it for yourself—if you have sovereignty over your food and the resources, especially decent land, to grow it—rather than having to buy it. If we want a world that is fed, we should focus not on feeding the world but on helping the world feed itself.

Much of what we grow in the United States does not become food anyway. As of 2022, 44 percent of the U.S. corn crop gets brewed into ethanol for the gas tanks of cars.[41] The situation is much the same with soybeans, with 42 percent of soybean oil going to concoct biodiesel for trucks.[42] There are some leftovers after turning corn into ethanol and soybean oil into biodiesel. These distiller's grains, as they're called, can be fed back to livestock, recovering some of their food value. But we could also feed much of our livestock on well-managed grass, growing and glowing across the landscape. And we could also try eating less meat.[43]

So why do we grow so much in places like Iowa? Because of how the treadmill of production and the farmer's problem promote growing way more than we can eat. Not because we need so much to eat.

But nonetheless, we do need to eat. Human production does have to come from somewhere, and there are a lot of us now. Do it wrong, and we mess up the world pretty badly. Climate change. Biodiversity loss. Soil loss. Water and air pollution. Dire inequalities in the impacts. We don't have to descend into Malthusian doomsday thinking, though. We can have an agricultural landscape that is "multifunctional"—a landscape that sustains us all, human and nonhuman alike, with more justice and deeper mutuality.

Iowa used to have such a landscape. Before the coming of the unrelenting productivism of post–World War II agricultural industrialism, there was still room for more than humans and for the mutual support the intertwined lives of all species provide one another. As Laura Jackson observed in her keynote address at the 2022 Practical Farmers of Iowa meeting, "There were many, many places in those early years—the teens and twenties—in which you could say this is an agricultural landscape and a prairie landscape at the same time." There was still space to appreciate, in Laura's words, "the majesty and the diversity of pre-invasion Iowa." Today that diversity is now confined to the few little remnants that have never been plowed. Like Rochester Cemetery in Rochester, Iowa: on those eighty acres, Laura said, one can still find more than four hundred species of plants alone.

"Would that be possible today?" she challenged the audience. Could we have that kind of balance on a wide scale again, and not just in a few unusual situations, like that cemetery? Could we have an agriculture that stores and restores carbon rather than pumping it up into the atmosphere? Could we have an agriculture that builds rather than loses soil? Could we have an agriculture that supports our communities, providing fair wages and maintaining crucial institutions like schools, health care, and the gathering places that deepen our ties with one another? Could we have an agriculture that provides healthy food without the poison, the dangerous machines, the pressure to compromise social justice and ecological well-being? Could we really have farming for us all?

For members of PFI and organizations like it across the United States and around the world, there is a growing appetite to farm more sustainably. They often use different vocabularies. Some describe what they do as agroecology (my own favored term). Others as regenerative agriculture. Still others as organic agriculture, renewable agriculture, food sovereignty, urban agriculture, permaculture, biodynamics, climate-smart agriculture, ecological farming, alternative agriculture, peasant agriculture, community agriculture, and more.[44] But these are all shades of a common green realization: we know we can farm differently because so many farmers already do.

We can eat differently, too, and millions already are. Some 140,000 U.S. farms now sell at least part of their yield directly to stores, restaurants, institutions, and households—$10.7 billion dollars' worth, as of 2020.[45] Most U.S. cities now have restaurants that serve farm-to-table fare. The number of farmers' markets in the United States has grown nearly fivefold in the last thirty years, from 1,775 in 1994 to 8,771 as of 2019.[46] (Madison, Wisconsin, the modest-sized city where I live, now has 15.) Some 10,000 farms operate a CSA for at least part of their sales.[47] One study tabulated over 18,000 community gardens in the United States.[48] Organic food claims 15 percent of fruit and vegetable sales in the United States as of 2022, and 6 percent of food sales overall.[49] Thirty-five percent of U.S. households raise at least some of their own food, whether in the back garden or on a balcony high up in an apartment building.[50] Gen Z and millennials show particular interest in the pleasures of food as local as you can get: from right outside your own kitchen.[51]

Something has changed. It wasn't like this twenty years ago, when the first edition of *Farming for Us All* appeared. A new tide is rising. It is rising in part through better government policies, like the Conservation Stewardship Program of the National Resources Conservation Service; the USDA organic certification program (which only dates from 1997); the nutrient management planning implemented by county, state, and provincial governments; and the increased commitment to sustainable practices by agricultural extension services. It is rising through better policies by institutions outside of government, including businesses, nonprofits, and universities that seek to encourage greener and fairer food through their food purchasing and provisioning, implementing standards like the Fair Food Program of the Coalition of Immokalee Workers, a worker-driven partnership that began among Florida tomato pickers and retailers. And it is rising from a sea change in university research and education, where words like "sustainable," "organic," and "agroecology" no longer need to be said in hushed tones lest the administrators at our colleges of agriculture overhear—and where students pack courses with these words in their titles.[52]

But what really underlies it all is that people now identify with a different ethic—an ethic of care for each other and the earth. I know that sounds rather hearts-and-flowers, a Mother's Day card of concern and respect. But we really should love our ultimate mother. Increasingly, we do.

That is the central point of this book: identity can overcome economics. Identity is both social and ecological. You've heard it before, but it really is true: it's all about relations. Yes, the deck is stacked against us. Yet we can cultivate different ways to be and thus different ways to know and do. Practical Farmers of Iowa has shown us that such a transformation is as possible as it is necessary. So have the thousands of other sustainable food and agriculture groups that have sprung up all across the United States and the world, like grasses greening back the land after a prairie fire.

In June 2023, I found myself in Iowa again, singing a song that had come to me as I was driving to that Practical Farmers of Iowa meeting the year before. My friends Jason and Ehler from the Barn Owl Band backed me up on fiddle and bouzouki as I sang these words at our concert in Ames, right in the center of the Des Moines Lobe, the fertile center of all that is Iowa and not the Ioway.

Acres and acres of corn
Riches I fear we must mourn
Green gold concealing the feeling of stealing

From those who have yet to be born
Acres and acres of corn

Acres and acres of beans
Feed to the mouths of machines
I know that they tell us and sell us its well for us
I think that's not all that it means
Acres and acres of beans

Acres and acres of yield
From every quilted green field
Fills me with sadness not gladness this madness
The iron and poison we wield
Acres and acres of yield

Acres and acres of cash
Seeded by greed and by gash
Think it will dry up and fry up and die up
When comes the climate's great crash
Acres and acres of cash

Acres and acres of hope
Restoring each gullied brown slope
It's not too late friends don't wait friends for fate friends
To tumble us from this tight rope
Acres and acres of hope

Acres and acres of hope. That's precisely what Practical Farmers of Iowa cultivates.

—MMB, Madison, Wisconsin, December 2023

1. Smiley ([1991] 1998, 16).

2. Median farm income was negative in 2016, 2017, 2018, and 2020 and only very slightly positive in 2019; see USDA Economic Research Service (2021) and the more detailed discussion later in the preface.

3. This term is not from the original edition of *Farming for Us All*. I am introducing it here. Closely related is the concept of "interests bias" that I discuss briefly in Bell (2018, 226).

4. Ashwood et al. (2014).

5. Here, I am quoting from Zach's presentation at PFI's online annual meeting in 2021.

6. The 83 percent figure for total rural poverty comes from the United Nations Development Programme and the Oxford Poverty and Human Development Initiative (2022), and it is based on the Multidimensional Poverty Index for the period 2016 to 2019. Using purely financial metrics, Castañeda et al. (2018) calculated that 80 percent of world poverty is rural, based on 2014 data. The figure for 43 percent of world population as rural is from the World Bank (2022).

7. Krugman (2022).

8. At this writing, in the summer of 2023, the results of the 2017 census are the most recent available.

9. It's actually fallen a bit since 2017, in part due to the COVID pandemic.

10. Leslie, Wypler, and Bell (2019).

11. U.S. Agricultural Census for 1920, page 295, table 2, from the USDA Census of Agriculture Historical Archive, https://agcensus.library.cornell.edu/census_year/1920-census/.

12. Race figures for Iowa farm producers from National Agricultural Statistics Service (n.d.), reporting on the 2017 agricultural census.

13. As well as the Southern Homestead Act of 1866.

14. Leslie, Wypler, and Bell (2019).

15. Crider (2021).

16. See Legal Services Corporation (2022) for the total number of U.S. "farm workers." Their figures are likely an underestimate, as many "farm workers" have uncertain legal status and thus are difficult to enumerate. On the percentage of foreign-born farm workers, see Hernandez and Gabbard (2018) on the results of the 2015–16 National Agricultural Workers Survey.

17. Janes Ugoretz (2023).

18. Hernandez and Gabbard (2018). Although 24 percent of "farm workers" identify as White, 83 percent identify as members of Hispanic groups, indicating that some identify as both Hispanic and White.

19. Janes Ugoretz (2023).

20. National Farmers Union, "About NFU," accessed June 19, 2023, at https://nfu.org/about/.

21. National Farmers Union, "Fairness for Farmers: A Farmers Union Project," accessed June 19, 2023, at https://nfu.org/fairness-for-farmers/.

22. A remnant of the Farmer-Labor Party exists to this day as the Minnesota Democratic-Farmer-Labor Party, now an affiliate of the U.S. Democratic Party. I thank my good colleague Nan Enstad for deepening my understanding of the history of who is considered to be a "farmer."

23. Iowa Department of Agriculture and Land Stewardship, "Century Farm Program / Heritage Farm Program," accessed June 19, 2023, at https://iowaagriculture.gov/century-and-heritage-farm-program.

24. Wise Voter (2023).

25. The adjacent states of Nebraska, South Dakota, Minnesota, and Wisconsin have many reservations, two of which—Nebraska's Omaha and Winnebago reservations—extend slightly into Iowa.

26. See U.S. Bureau of the Census (1984), table 1.

27. National Agricultural Statistics Service (2021).

28. Cochrane (1958).

29. USDA Economic Research Service (2023b).

30. Carr and Schechinger (2021).

31. Schnaiberg (1980).

32. USDA Economic Research Service (2022c). To be sure, some of this low farm income may be because of accounting practices that enable farmers to hide their income from the IRS.

33. Legun and Bell (2016).

34. Technically, it is the lowest ever recorded, as we only have this data back until 1993. USDA Economic Research Service (2022d).

35. I am quoting "The Farmer Is the Man," which I sometimes sing as "The Farmer Is the One."

36. American Farm Bureau Foundation for Agriculture (2021).

37. Schechinger and Cox (2016).

38. U.S. Government Accountability Office (2023).

39. USDA Economic Research Service (2023b).

40. Ibid.

41. USDA Economic Research Service (2023a).

42. USDA Economic Research Service (2022b).

43. Mayerfeld (2023).

44. Newton et al. (2020).

45. USDA Economic Research Service (2022a).

46. Ibid.

47. Numbers vary widely, though, depending on the methods used to tabulate the number of CSAs.

48. This is a commonly cited figure on the web, attributed to the American Community Gardening Association at sources such as this one from the Boston Public Library: https://guides.bpl.org/communitygardening. I regard the figure as somewhat squishy, though.

49. German (2023).

50. Mayers (2023).

51. Ibid.

52. I can report that these really were words that could get you into trouble in our colleges of agriculture twenty years ago—and even ten years ago. In some situations, they still do, but that is becoming less common.

REFERENCES FOR THE PREFACE

Alkire, Sabina, Mihika Chatterjee, Adriana Conconi, Suman Seth, and Ana Vaz. 2014. *Poverty in Rural and Urban Areas: Direct Comparisons Using the Global MPI 2014*. Oxford, U.K.: Oxford Poverty and Human Development Initiative.

American Farm Bureau Foundation for Agriculture. 2021. *Food and Farm Facts*. Washington, D.C.: American Farm Bureau Federation.

Ashwood, Loka, Noelle Harden, Michael M. Bell, and William Bland. 2014. "Linked and Situated: Grounded Knowledge." *Rural Sociology* 79 (4): 427–52.

Bell, Michael M. 2018. *City of the Good: Nature, Religion, and the Ancient Search for What Is Right*. Princeton: Princeton University Press.

Carr, Donald, and Anne Schechinger. 2021. "Do Farmers Need More Federal Welfare?" Environmental Working Group. Accessed June 29, 2023, at https://www.ewg.org/.

Castañeda, Andrés, Dung Doan, David Newhouse, Minh Cong Nguyen, Hiroki Uematsu, João Pedro Azevedo, World Bank Data for Goals Group. 2018. "A New Profile of the Global Poor." *World Development* 101: 250–267.

Cochrane, W. 1958. *Farm Prices: Myth and Reality.* Minneapolis: University of Minnesota Press.

Crider, Margaux. 2021. "'The Entire Creation Is Within Them': Gender, Ecology, and *Viriditas* as Lived Religion." Master's thesis, University of Wisconsin–Madison.

German, Brian. 2023. "Report Shows Organic Food Sales Hit a New Milestone in 2022." AgNet West Radio Network. https://agnetwest.com/report-shows-organic-food-sales-hit-a-new-milestone-in-2022/.

Hernandez, Trish, and Susan Gabbard. 2018. *Findings from the National Agricultural Workers Survey (NAWS) 2015–2016: A Demographic and Employment Profile of United States Farmworkers.* Research Report 13. JBS International. Report prepared for the U.S. Department of Labor, Employment and Training Administration, Office of Policy Development and Research.

Janes Ugoretz, Sarah M. 2023. "Supporting the Human Element of Farming: Building Long-Term Careers on Small-Scale Vegetable Farms Through Increased Social Sustainability." Ph.D. diss., University of Wisconsin–Madison.

Krugman, P. 2022. "Wonking Out: Facts, Feelings, and Rural Politics." *New York Times,* Oct. 21. https://www.nytimes.com/2022/10/21/opinion/rural-america-politics.html.

Legal Services Corporation. 2022. "Summary Table of National and State Estimates of the 2021 LSC Agricultural Worker Poverty Population (1): Final Estimates." https://lsc-live.app.box.com/s/9zb1ak1knkzabrohs1mstuoeckcd9i06.

Legun, Katharine, and Michael M. Bell. 2016. "The Second Middle: Conducers and the Agrifood Economy." *Journal of Rural Studies* 48:104–14.

Leslie, Isaac Sohn, Jaclyn Wypler, and Michael M. Bell. 2019. "Relational Agriculture: Gender, Sexuality, and Sustainability in U.S. Farming." *Society and Natural Resources* 32 (8): 853–74.

Mayerfeld, Diane, ed. 2023. *Our Carbon Hoofprint: The Complex Relationship Between Meat and Climate.* Cham, Switzerland: Springer.

Mayers, K. 2023. "Gardening Statistics in 2023 (incl. Covid & Millennials)." Garden Pals. Accessed June 29, 2023, at https://gardenpals.com/gardening-statistics/.

National Agricultural Statistics Service. 2021. *Farms and Land in Farms, by Sales Class—Iowa: 2016–2020.* Accessed June 29, 2023, at https://www.nass.usda.gov.

———. N.d. *Iowa 2017 Agricultural Census Racial Profile.* Accessed August 6, 2023, at https://www.nass.usda.gov.

Newton, P., N. Civita, L. Frankel-Goldwater, K. Bartel, and C. Johns. 2020. "What Is Regenerative Agriculture? A Review of Scholar and Practitioner Definitions Based on Processes and Outcomes." *Frontiers in Sustainable Food Systems* 4. doi: 10.3389/fsufs.2020.577723.

PFI (Practical Farmers of Iowa). 2022. *2021 Annual Report.* Ames, IA: Practical Farmers of Iowa. Accessed Aug. 8, 2022, at https://practicalfarmers.org.

Schechinger, A. W., and C. Cox. 2016. *Feeding the World: Think U.S. Agriculture Will End World Hunger? Think Again.* Washington, D.C.: Environmental Working Group.

Schnaiberg, Allan. 1980. *The Environment, from Surplus to Scarcity.* New York: Oxford University Press.

Smiley, Jane. [1991] 1998. *A Thousand Acres.* London: Flamingo.

United Nations Development Programme and Oxford Poverty and Human Development Initiative. 2022. *Global Multidimensional Poverty Index 2022: Unpacking Deprivation Bundles to Reduce Multidimensional Poverty*. Oxford, U.K.: United Nations Development Programme and Oxford Poverty and Human Development Initiative.

U.S. Bureau of the Census. 1984. *1982 Census of Agriculture*. Vol. 1, *Geographical Area Series. Part 15: Iowa State and County Data*. Washington, DC: U.S. Government Printing Office.

USDA Economic Research Service. 2021. "Farm Household Well-Being: Farm Household Income Estimates." Accessed Aug. 6, 2022, at https://www.ers.usda.gov.

———. 2022a. "Charts of Note: Growth in the Number of U.S. Farmers Markets Slows in Recent Years." Accessed June 29, 2023, at https://www.ers.usda.gov.

———. 2022b. "Examining Record Soybean Oil Prices in 2021–22." Accessed June 29, 2023, at https://www.ers.usda.gov.

———. 2022c. "Farm Household Well-Being: Farm Household Income Estimates." Accessed June 29, 2023, at https://www.ers.usda.gov.

———. 2022d. "Farm Share of U.S. Food Dollar Reached Historic Low in 2021." Accessed June 29, 2023, at https://www.ers.usda.gov.

———. 2023a. "Feed Grains Sector at a Glance." Accessed June 29, 2023, at https://www.ers.usda.gov/topics/crops/corn-and-other-feed-grains/feed-grains-sector-at-a-glance/.

———. 2023b. "Government Payments, Farm Income and Wealth Statistics." Accessed June 29, 2023, at https://www.ers.usda.gov.

U.S. Government Accountability Office. 2023. "International Food Assistance." Accessed June 29, 2023, at https://www.gao.gov/international-food-assistance.

Wise Voter. 2023. "Native American Population by State." Accessed Aug. 7, 2023, at https://wisevoter.com/state-rankings/native-american-population-by-state/.

World Bank. 2022. "Rural Population." *The World Bank: Data*. Accessed Aug. 6, 2023, at https://data.worldbank.org/indicator/SP.RUR.TOTL.ZS.

OVERTURE: *Cultivating Sustainability*

"Do you want your neighbor or your neighbor's farm?"

Dick Thompson is in his machine shed, microphone in hand, striding across the front of an impromptu auditorium of folding chairs occupied by some one hundred farmers, university people, and others among the agriculturally curious. With the portable speaker slung over his shoulder Dick can move around a lot, and he does, asking for questions and answering questions, asking for answers and questioning answers: a talk-show host in blue overalls. He's wearing Liberty overalls today, and it says so in blue letters across the front of the bib. He is also wearing his trademark red shirt (sleeves rolled up) with black wing-tips on his feet, gold-rimmed glasses on his blue-eyed ruddy face, and a blue cap on his silver-white hair that reads "PFI: Practical Farmers of Iowa." It's the annual PFI field day on the Thompson farm. Dick is explaining how it is that he, Sharon, their son Rex, and their daughter-in-law Lisa can make a good living for two families in an environmentally friendly way on just three hundred acres, while most of the rest of Iowa's farmers are busily gobbling up their neighbors' land, and thus their communities, in order to reach the magical goal of *A Thousand Acres*—to quote the title of Jane Smiley's dark novel of Iowa agriculture. And every year a lot of people come to the Thompson farm, and to other PFI farms, to find out just how they manage it.

"Your neighbor or your neighbor's farm?" Dick repeats. "You've got to ask yourself that."[1]

He lets the point settle in, and the crowd thinks over the seemingly inexorable advance of agricultural industrialization and modernization across the American rural landscape that each year drives out another half percent or so of farms. Most farmers in Iowa rely on their corn and soybean crops, and 60 percent of the state is covered by these two plants alone, some 22 million out of the state's total of 36 million acres.[2] Prices vary, but in a good year a grain farmer can expect to make maybe $30 to $40 an acre in profit—but only after the government chips in $30 to $40 in subsidies. When the federal government is feeling particularly generous, as it has been since 1999, roughly doubling subsidies over previous levels, that figure can rise to $60 to $70 an acre. Increasingly, what makes for a "good year" is not the climate in Iowa but the climate in Washington, D.C.[3]

Which means several things. It means that without government subsidies, the average Iowa farm, as currently managed, would be broke.[4] It means that if you're an Iowa farmer and you want to attain a mid-level income from your farming, say $40,000 to $60,000, which are roughly the median figures for U.S. households of more than one person, you need something very close to Jane Smiley's

Dick and Sharon Thompson at their farm in Boone, Iowa, 1999.

thousand acres of farmland—and the industrial farming machinery and industrial farming approach that make it possible to wrest crops out of so much ground.[5] It also means that there will be tremendous pressure to increase your farm size by whatever means possible, because Iowa, as of 2002, had 32.7 million acres of farmland and about 92,500 farms.[6] That's 354 acres apiece, leaving the average Iowa farmer 646 acres, or almost two neighbors' farms, short. In fact, something like two or three thousand acres apiece would be better, say many, because some years you're lucky to make $10 an acre, even with government subsidies. That's eight neighbors' farms short—and, consequently, eight former neighbors. Under conditions like these, it's hard to pay much attention to the disappearing soil, the disappearing wildlife, and the disappearing community life that the Big Farm, Big Tractor, and Big Chemical way sends down the creek.

The Thompsons, however, get by with even fewer than 354 acres, and for not one but *two* families. By having a small farm, the Thompsons are able to manage each acre with exceptional care, minimizing reliance on the surefire chemistry of Monsanto and Dupont and thus minimizing cost and environmental damage as well. It's not an organic farm, but they have used pesticides only once in the past twenty years. They use no antibiotics or hormones in their pigs and cattle, they do not plant genetically engineered crop varieties, and yet they have some of the highest crop yields and lowest soil erosion rates in their county. They also have a solid, though not lavish, farm income—without government subsidies, having long ago sworn to refuse them, an act of defiance that many find particularly confounding.

And they have plenty of that well-known source of the farmer's pride: "green paint"—Iowa rural slang for farm equipment, from the distinctive color of Iowa's most popular brand, John Deere. It's the Thompsons' favorite brand, too. They have three John Deere tractors, all complete with air-conditioned cabs. They also have some farm equipment that most people don't have, like a rotary hoe for controlling weeds without chemicals, a Buffalo planter for "ridge tilling" (one of the techniques the Thompsons and some others use for limiting chemicals and conserving soil), and a tiny combine that does only four rows of crop at a time but allows the Thompsons to monitor their fields better. In fact, all the Thompsons' machinery is "four-row equipment," well below the eight-, twelve-, sixteen-, thirty-two-, and sometimes even sixty-four-row equipment to be found on most other Iowa grain farms. The Thompsons are constantly running their own experiments on their farm, and the four-row equipment is a big help. Call it "precision farming," but of a much different type from the computerized, satellite-linked, global-positioning-system farming that turns the tractor (and perhaps the farmer) into something of a robot, which is what the term "precision farming" usually means in farming circles. Not that the Thompsons are anti–high-tech. Far from it. For example, a couple of years ago they experimented with planting in the dark using an infrared viewing helmet, to see if weed seeds stirred up in planting were less likely to germinate without the stimulus of light. (They weren't, on the whole.) Posters on the results of their years of experiments are on display all around the machine shed, as is most of their equipment—including the Buffalo planter, which during the years of their night-planting experiments they draped in old carpet, a technique they cooked up to keep even the moonlight out of the furrow.[7] Some of it may be unusual equipment and the Thompsons may do unusual things with it. But they clearly can afford to buy it, and that is culturally very persuasive among farmers. Moreover, they buy only in cash and carry no debt load on their farm, another radical departure from conventional practices on Iowa farms.

Dick continues. "The problem is, we're raising commodities out here, not crops. But commodities don't make communities. It takes people to make communities."

For anyone who has grown up in rural Iowa and watched one local institution after another shut down and crumble away over the years, this hits particularly close to home. As prosperous and tidy as the horizon-to-horizon grain fields may look, the Main Streets of many Iowa small towns resemble long-abandoned movie sets. It is not unusual to find half the shop fronts dustily vacant or grimly boarded up. Empty stores, empty schools, empty churches, empty hospitals: the economic situation is no worse in Watts or the South Bronx. Out in the countryside, away from the exurban aura of Iowa's few major cities, abandoned houses and farmsteads stare at the passing motorist like skulls bleaching in the sunlight, their paint peeling down to the bare bone of the wood and their glassless windows vacant as empty eye sockets. (And this is despite the constant work of pulling rural buildings down, cleaning up the evidence of decline.) Writing during the farm crisis of the 1980s, author Osha Gray Davidson called it the "rural ghetto," and things have not gotten much better since.[8] "Abandoned Iowa" one could also call it, a jarring image to hold in one's mind alongside that of the $200,000 combines and $100,000 tractors working the elegant lines of nearly weed-free crops—too jarring for many, and so they drive right on by, keeping their focus elsewhere on the passing landscape.

It was in 1985, during the middle of the 1980s farm crisis, that a group of Iowa farmers and farm advocates found that they could no longer close their mind's eye to the decline, the abandonment, and the environmental degradation, and started Practical Farmers of Iowa, with Dick Thompson as the first president. Convinced that it must be possible to farm in economically and environmentally sound ways on small farms that support community life, PFI's founders dedicated the group to sharing information among farmers about how it could be done and encouraging farmers to conduct and disseminate the results of their own on-farm research. University researchers at the time were paying next to no attention to anything other than the industrial model of farming, and very little relevant research about other approaches was available. So PFI organized field trials that would be statistically valid, using randomized and replicated plots that could be subjected to tests of significance and other statistical techniques. The local land grant university, Iowa State, shortly afterward took interest in PFI's embrace of a scientific approach and agreed to form a highly unusual partnership with the organization, giving it a university office and giving its few staff members (who numbered only one at the time) the status of university employees, although PFI provided the funds for their salaries. By the early 2000s, the group had grown to some seven hundred member

Abandoned farmstead near Webster City, Hamilton County, 2000.

households, about half of whom farm. And now a couple of dozen faculty and researchers at Iowa State regularly work with PFI farmers, and many of them are members of the group.

A lot has been learned since 1985. What has come to be called "sustainable agriculture" has become a major focus of interest not just for farmers but now also for researchers, policy makers, and consumers—and not just in Iowa but across the world.[9] There are hundreds if not thousands of sustainable agriculture organizations today, some governmental, some nongovernmental, and some, like PFI, a bit of both. (Iowa State University, which is an institutional partner of PFI, is state supported.) In Iowa, advocates of sustainable agriculture have been particularly excited by new farming techniques like ridge tilling, rotational intensive grazing, deep-bedded hoop houses for hogs, and holistic management, as well as older techniques like crop rotation, flame cultivation, pasture-farrowing, and direct marketing—jargon to those outside what must now be recognized as the *sustainable agriculture movement* but a social, economic, and environmental lifeline for those inside it.[10]

A lot has been learned and many are involved, but it would be safe to say that sustainable agriculture remains only a minor player in the global agricultural system. The vast bulk of research at places like Iowa State University still follows the industrial mentality. A lot of basic research on sustainable agriculture has yet to be even begun.

Still, Dick says to the audience, perhaps with a whisker of overconfidence,

"Some of us know what to do. The question is, will we do it?" Dick has a way of challenging his listeners that most rise to, and today's audience is no exception.

"So why *won't* we do it?" a middle-aged farmer in the group calls out.

Dick breaks into a big smile. He's been waiting for this question. But he holds his own views back a bit. "Well, what do you think?" Dick returns. "What are some of the reasons?"

"Communication!" calls out one voice.

"Education. It's what's in the farm magazines. Or rather what isn't in them!" calls out another.

"Ya, ya. That's part of it," Dick encourages. "But there's more to it. Keep going. Keep going."

"Cultivation, I guess," another farmer responds after a moment, meaning mechanical hoeing, one of the main ways to control weeds without chemicals. "Farmers today don't like to cultivate. It takes too long. And they don't like livestock either. That takes too much time, too."

"That's part of it, too," Dick affirms. "We're getting closer."

A hand goes up in the front row. "The problem is, the average farmer today wants to have twelve months' income with two months' work. A month of planting and a month of harvesting."

This is something of an exaggeration, but the words strike a chord with an audience that is well aware of the commonly held first-one-to-Florida ideal of farming. And it is true that sustainable farming methods often require substituting more labor time and more management time per acre in place of big capital outlays for big technology and the big land base needed to pay for it. That's all right, say most sustainable farmers, because labor and management skills are our strengths, so we don't need big capital, big technology, and big land. But it does run counter to the industrial image of what progressive farming is about to invest so much time and effort per acre doing things like cultivating, when an herbicide laid down with the seed might take care of the weeds from the start.

As for raising livestock, this is one of the main means sustainable farmers have for making a go of it without grasping for the neighbor's land. Livestock provide a farm with nonchemical fertilizer (manure, that is), a way to make money from hay during the crop rotations needed to break pest cycles without chemical pesticides (you can feed the hay to your livestock), and "value-added" products like meat and milk that bring a little more home than "two-dollar corn" (corn selling in the range of $2.00 to $2.99 a bushel, the price farmers have seen most of the last ten years) or even "one-dollar corn" (corn selling in the range of $1.00 to $1.99, a shockingly low level that has started showing up regularly in the last few years). Until recently, a good deal more than half of all Iowa farms raised some form of livestock.[11] But

The landscape of industrial agriculture: grain wagon, grain elevators, and combine, near Gowrie, Iowa, 1989.

many farmers have been moving into growing grain alone, in part because livestock markets have been highly volatile of late, in part because livestock tie you to the farm year-round, and in part because of the rise of contract livestock production, which has increased the scale of many operations and limited access to markets for the remaining smaller ones. The trick in the sustainable approach is to recognize that livestock can be more than simple cash cows—that livestock can be managed so as to enhance the profitability and environmental friendliness of the whole farm operation, even if they do take more labor and management time. Besides, if there are more farms around, it will be that much easier to find a neighbor to do the feeding when the family does go off to Florida for winter vacations.

"You're getting close," says Dick to the audience. "Getting real close."

He looks over the crowd in the machine shed, and the crowd looks back. The time has come to put it all together.

"Greed and ease. That's it," Dick breathes into the microphone, and a lot of heads give a slow nod of recognition and agreement at this critique of the materialist ambitions of industrial agriculture. Knowing glances are exchanged. "The other way's easier."

"But is it?" asks the farmer who had earlier pointed to the problem of what is and is not in the farm magazines. The mood in the shed is crackling now, and Dick doesn't need to ask for input. Hands are going up everywhere and a couple of people are standing. "I mean, you do it, and you seem to live well. And your neighbors must see that. So what do they think? Do they ever ask you how you do it?"

Even Dick pauses at this one, and the whole shed pauses with him. "I call that a social problem," he begins. "I guess that's just the way most guys are. I hardly ever listened to my father, at least when he was alive." Dick in fact now farms quite a bit like his father did, using a variation on the five-year crop rotation his father developed in the thirties and forties and hardly ever using farm chemicals. "You don't listen to those close to you, it seems. Maybe it takes a farmer in the next county doing something to show you."

Dick straightens up a bit and flashes his wide smile. He takes the portable microphone up to his mouth with both hands, and adds the kicker. "But the main issue for all of us is this: Do I really want to know? And if I do know, do I want to do anything about it?"

Or, to put Dick's question more generally, why don't more farmers change? The current situation for most farmers in Iowa, as elsewhere in the United States, is one of uncertainty propped up with doubt and risk—of fierce struggle on a playing field tilted steeply by the structure of regulations and government subsidies, as secure as the next national election. And down that steep incline tumbles a continual detritus of former farmers, boarded-up farm towns, and the soil itself, kicked up and washed away by the scuffle of the play. Except that it's not a game: We're talking about real lives, those of farmers and us who eat, for what happens to farmers has implications for everyone, and for the life of the land that must sustain us.

But is the situation really so desperate? There is a statistic that, if you live in Iowa, you are bound to encounter eventually: "One Iowa farmer feeds 220 people." It's the kind of number that finds its way into the keynote address at the annual dinner of the local county chapter of the Iowa Corn Producers Association, and into the tourist brochures you can pick up in rest stops along Iowa's sections of Interstate 80 and Interstate 35. One could quibble with how such a figure is tabulated. Iowa's agriculture concentrates almost entirely on two things, meat and the feed grains used to produce it, and couldn't be said to completely round out the total diet of very many people at all. But it's not the kind of statistic that anyone, aside from statisticians, needs to question closely.[12] We all know what it is trying to say: what a blessing the industrialization of agriculture supposedly has been. Never have so few fed so many, opening up economic space for the wide range of nonagricultural livelihoods that most of us now pursue and the benefits that those livelihoods bring.

A couple of other statistics (ones that are considerably more valid) make much the same point. The time is not so long distant when all but a few percent of the world's human population gained at least a significant portion of their living from farming. Today in the industrialized countries the figure is reversed. In the United States, only 2.3 percent of the labor force is employed in farming.[13] Similar figures apply throughout most of Europe as well as to some countries of the Far East, such as Japan and Taiwan. If people weren't free to work in something other than farming, we could not easily have schools, computers, hospitals, the arts, ready-to-wear clothes, and more—delights and comforts and enrichments that few who enjoy them would gladly give up. This is undeniable, most agree, and indeed ranks as one of the touchstones of our cultural identity as modern people.

But those in the sustainable agriculture movement contend that the blessing of industrial agriculture has been decidedly mixed, pointing to the environmental problems, questions about food safety and consumer trust, issues of economic justice, community decline, and the threat to rural culture of the ever-increasing scale of farms and farm equipment. For the most part, advocates of sustainable agriculture cannot be described as antitechnology, though. Their argument is not with tractors and computers. Almost all sustainable farmers own at least one of the former and the latter. Nor are most sustainable agriculture advocates anti-schools, anti-hospitals, or anti–ready-to-wear clothes. A few home school their children, many are interested in traditional forms of medicine, and virtually all of them admire hand-knitted sweaters and other homemade clothes. But they are not calling for closing the local schools, hospitals, and stores—just the opposite, in fact.

The argument behind sustainable agriculture is not that we need to return everyone to the land. Rather, it is that the technological, economic, and political structures that consolidate farms eventually reach a point where they do more harm than good. It may be that only a few today need farm, but do they farm for us all? Do they farm for the health and well-being of consumers, rural communities, urban communities, even farmers themselves, as well as consumers and communities and farmers in other countries? Do they farm for the health and well-being of the environment? Sustainable agriculture advocates answer: on the whole, no.[14]

Moreover, say these advocates, there are environmentally and economically sound alternatives that can slow the tide of farm loss, maybe even reverse it somewhat, without threatening either our labor supply or our food security. After all, if only a few percent of working adults are farmers now, agriculture hardly represents a significant reserve pool of workers for other sectors of the economy anymore.

Auctioneers and bidders for the furnishings of Rolfe Presbyterian Church, razed when the town's Methodists and Presbyterians formed the Rolfe Shared Ministries in response to low numbers of parishioners, 1996.

And an approach to productive farming that encourages greater stewardship of the environment's productive potential should increase food security, not decrease it. Such an approach is *technically* possible, say the advocates of sustainable agriculture. The real issue is how it can be *socially* possible.

Which is the question I address in this book, with the help of my colleagues Sue Jarnagin, Greg Peter, and Donna Bauer. For ten years, from 1994 to 2003, we investigated the social soils of agricultural sustainability through a study of one sustainable agriculture organization—PFI, Practical Farmers of Iowa—and its human landscape. Sue, Greg, Donna, and I conducted interviews, attended meetings, rode tractors, pitched hay, helped fix broken equipment, ate, washed up, played volleyball, drank beer, chased children, and just plain "visited," in the Midwestern sense, with several dozen PFI farm households, several dozen of their neighbors, and several dozen more of the group's nonfarming associates in the university and elsewhere. We did so in the hope that we might thereby gain an understanding of the culture of sustainable agriculture and of the cultivation of the sustainable agriculture movement—culture and cultivation in the social sense. For agriculture has always been more than a matter of agronomics alone.

And we also did so because we believe that sustainable agriculture is a social cultivation of great significance—to rural life, to urban life, to the environment, and, as will emerge later, to fundamental issues of knowledge and democracy.

Under some conceptions of social science, such an admission is at best a major tactical blunder and at worst clear evidence of biased scholarship. But I write this book with the thought that social scientists do not have to ignore or squelch their own values in order to conduct meaningful research. Indeed, the reality is that social scientists cannot ignore or squelch their own values—cannot both in the sense of being unable to and in the sense of its being inappropriate to. The most value-*free* social science may well turn out to be the most value-*less*. Rather than being a problem for social research, values are the whole reason for social research. Of course, social research that is value-*laden* is likely not to be valu-*able*. But the problem of research that is value-laden is not the presence of values in it. Rather, it is that the values are unacknowledged and their implications for the research are not carefully considered. What is needed is for social researchers to be aware of their values and to communicate them to others, so that both researcher and reader can take them into account in formulating and evaluating a study and its arguments.

The acknowledgment of values is also crucial if researchers are to avoid the tendency to ignore uncomfortable evidence and arguments that may contradict their own. Consideration of alternative interpretations that others have pointed out, or might well point out, is the hallmark of rigorous scholarship. Being up front with oneself and with others about one's values should help ensure that one's research does take alternative interpretations into consideration, offers reasons for disagreement, and provides others with access to that reasoning and the evidence upon which it is based. It's a matter of responsible scholarship—that is, scholarship that encourages further response to its own interpretations through its consideration, reasoned engagement, and openness.

So let it be said now: This book is a study, rigorous and responsible, I trust, of something that I believe in.[15] My hope is that my own personal commitments to sustainable agriculture, and those of my colleagues, Sue, Greg, and Donna, have sharpened my critical sensitivities in the way that one can be most critical of those one holds most dear, and not merely dulled my sensitivities in the way that one can easily overlook the faults of close friends and family. That will be the reader's own responsibility to judge.

So why don't more farmers change? If we are to understand how a more sustainable agriculture might be socially possible, we must answer this question, the question that underlay the conversation in Dick Thompson's machine shed that fall day. Or, to turn the question around, why *do* some change—why do some farmers switch to sustainable agriculture? This is an equally central issue to sustainable agriculture's social possibility.

Answering this question both ways around is what I'll be spending the rest of the book doing. But let me spend a few pages now giving an overview for the reader on the run. Inevitably, compressing the argument of an entire book into a few pages makes for somewhat thick and perhaps tedious reading, especially for those less familiar with the goals and manners of academic sociology. Readers who find themselves getting bogged down in the rest of this chapter are hereby cordially invited to skip ahead to Chapter 1, to hear immediately from the farmers themselves. Don't worry about missing anything. You'll still get the whole argument, but in less abstracted—and less abstract—form.

But before anyone skips ahead, let me try my hand at summarizing the book's argument in one sentence, to better orient all readers to my purposes. PFI helps make sustainable agriculture socially possible by guiding and encouraging farmers in being better at talking with others—in engaging in open, critical dialogue—and thus better at the practical matter of getting things done in the world. That's all.

I take up the question of why more farmers don't change to sustainable agriculture in the first two sections of the book. One answer is the agricultural version of the oft-heard statement "people don't like change"—that is, that *farmers* don't like change. There are several problems with such a view. Farmers, like most people, are quite eager to inform a willing listener of the things they wish were better in their lives. Pretty much everyone, I think it safe to say, has desires, plans, ambitions, and dreams. What are these but changes people hope for and look to? So farmers, and others, often do like change.

Of course, there are many changes at work in our lives that we are not happy about, and generally these are changes that thwart our desires, plans, ambitions, and dreams. Indeed, it is generally because of the changes we don't like that we seek the changes we would like. And all farmers today experience change, as they try to ride the constantly transforming machinery of technology, markets, and regulation, as well as environmental transformations such as soil erosion, water shortages, and pest problems. So while switching to sustainable agriculture would entail significant changes for a farmer, not switching to sustainable agriculture is no way to avoid change. Not switching to sustainable agriculture is thus unlikely to be a matter of a farmer simply not liking change.

Another common answer to the question of why more farmers don't take up sustainable agriculture is that the structure of agriculture—that is, the technology, market forces, and government regulations that affect agriculture—makes it very hard to change to sustainable practices. Plus, the current structure gives farmers still in the game very little incentive to change. Sure, it might be hard on those who have been forced off the land. But if you're still farming, the argument goes, then the structure of agriculture must be doing okay for you. Moreover, those who are still "in," as farmers say, can quietly celebrate their survival as a sure sign that they are doing something right. Surely they must be more efficient, and working harder, than those who have failed.[16]

Farm structure is indeed important, very important, for understanding why more farmers don't switch to sustainable practices. But it's not the whole story. The situation of farmers is one of great uncertainty, even for those who are still "in," at least in Iowa. Acquiescing to the structure of agriculture does not relieve a farmer of pressures, doubts, and troubles. Most Iowa farmers experience continual economic crisis, and all are constantly aware of the economic sword dangling but a few inches above them. "The fun's gone out of it now," is a common refrain among Iowa's farmers. Few can be said to be clearly doing well. The same can be said of their emptying communities, and sometimes of their families too, stressed by the knowledge of the sword above and the time spent dodging and ducking as it swings in the economic and political breeze.

And given the dominant role of subsidies in contemporary agriculture, and the similarity of farm practices from Iowa grain farm to Iowa grain farm, it is not clear that a difference in efficiency has all that much to do with most farmers' ability to stay "in"—except efficiency in landing the subsidies.[17] Some Iowa farmers are remarkably efficient in this regard, taking in annual subsidies in the hundreds of thousands of dollars, while some are only getting subsidies of a few hundred dollars a year.[18] Many adept and hard-working farmers—farmers whose crops are not the ones favored by the current subsidies, whose families were not able to give them as much of an economic boost when they started, or whose farms experience a couple of unlucky years of weather—will find themselves without the same cushion and without the same ability to compete for land base as those with big subsidy dollars.

While it is true that the current structure of agriculture encourages farmers to farm the subsidies and not the land, many if not most Iowa farmers (and their communities, their environment, and their families) find themselves poorly served by this encouragement, I argue in the first section of the book.

In the second section, I argue that the reasons why most farmers nonetheless don't change to sustainable agriculture lie in matters of knowledge and its relationship to identity. Central to making it in farming are various recipes and routines of agricultural knowledge, ordered understandings of how to do what needs to be done. Farmers must contend with an endlessly perplexing tangle of interactive and changing factors: crop varieties, crop pests, soil fertility, markets, regulations, human tastes and values, equipment, buildings, access to land, access to capital, family situations, labor availability, off-farm work—and, of course, the weather. For most Iowa farms, this complexity has to be coordinated for multiple crops, and often livestock too. So there isn't time to continually reinvent and experiment. The corn has to be in the ground now, if there is to be time to plant your soybeans. The weeds need to be controlled now, if there is to be time to get to that second farm that you have just rented or that off-farm job you have just taken on to add a bit of income. The hogs need to be taken to market now, despite today's low price, if there is to be time tomorrow to get to your daughter's high school basketball game. Besides, you've missed the last three.

In other words, like most large endeavors, farming requires the acquisition of a vast array of tricks of the trade—some tricks you buy, like better equipment, better seeds, and better marketing software; some tricks you learn, like crop rotation, the peculiarities of your own farm, and the neighbors you can best trust to help out in a pinch; and some tricks you both buy and learn, like ridge tilling your corn and soybeans, a technique that requires both different equipment and different knowledge. Once acquired, you can't take the time to continually question the stock of tricks you have at hand. You have to take them for granted, because the wind is up and it is about to start raining, hard. Even when you have reason to believe that the knowledge you have doesn't really work, you go with it anyway, because at least it's *something* you can do. When the storm bursts, you seek the shelter at hand, even if you have reason to suspect the soundness of the roof.

This *phenomenology of farming*—this taking for granted of what you know works, even when you think it might not—is a matter of more than material and temporal investments, though. It is equally a matter of identity, of the investment of the self as a man, as a woman, as a farmer. What you know is who you are. The stocks of knowledge we each hold within are stocks of self as well. I am a sociologist not just because my brain retains sociological knowledge but because I identify with that knowledge, and others identify that knowledge with me. Farmers are farmers because they identify themselves with the knowledge of farming and others identify that knowledge with them. Increasing the refraction on the microscope, farmers are types of farmers—grain farmers, livestock farmers, industrial farmers, sustainable farmers—because of what they know, therefore do, and therefore identify with.

Plus, most of what we know is known, at least to some degree, by others, and most of what we learn is learned, at least to some degree, from others. You can't learn everything on your own. We take it from others that certain species of mushrooms are poisonous. We take it from others that there are machines and chemicals—computers, tractors, fertilizers, drugs—that can help us along through our day. Or perhaps we may take it from others that certain machines and chemicals may in fact hinder us. We don't have time to do all the experiments ourselves, to build the machines ourselves, to concoct the drugs ourselves. So we learn from, and with, others, and gain a sense of social connection thereby. Knowledge is cultivated within culture and culture's lines of difference and identity—what I will be terming the *cultivation of knowledge*. We know, therefore we are. Knowledge has a history, a social history, and we connect ourselves to that social history—and the social present and social future it implies—when we connect ourselves to knowledge.[19]

For this reason we are rarely content to ask only the question *what* knowledge. We also almost always ask *whose* knowledge. We want to know the social history of knowledge so we can locate it within social life. Did this knowledge come from the local farm chemical dealer? From the guys at the co-op? From the university extension agent? From personal experience? Each of these histories places knowledge within particular social settings, which in turn strongly influence how we regard it and act on it.

In other words, knowledge is a social relation. Knowledge is people. It ties us to some, and often disconnects us from still others. What you know is who your friends and *relations* are. And with identification with knowledge comes a sense of trust in it and those we received the knowledge from—the trust required to take that knowledge for granted when the wind is up and a hard rain threatens.

Which means that to give up a cultivation of knowledge, to give up a field of knowing and relating, is to give up both a field of self and its social affiliations and a field of trust in the secure workings of the world. That's a lot to give up.

But some farmers do change to sustainable practices, despite the structure and phenomenology of agriculture. In the third section of the book I take up this question, "visiting with," as Iowans say, the farmers and farm families of Practical Farmers of Iowa.

What PFI members say is that there are times when one's trust in a cultivation of agricultural knowledge—that nexus of identity and the taken-for-granted—is called into deep question. Other things being equal, retaining a cultivation yields a steady crop of trusting comfort in what you know and who you are. But sometimes other things are not equal. Sometimes that trusting comfort erodes away under the force of a steady rain, a rain of suspicion that other people's social interests have engineered the crops of knowledge in one's own fields.

Farmers are not unaware that knowledge and interests are connected.[20] They routinely distinguish, for example, between the knowledge cultivated by university scientists, the extension service, government agencies, agribusiness chemical and implement dealers, farm commodity groups, and the chat by the coffee pot at the local grain elevator. They know that there are different interests involved in each case. They may not always stop to consider with care the interests behind each knowledge claim. Indeed, frequently they don't. And they may not always recognize what those interests are, even when they do stop to consider the matter. Again, frequently they don't. But they are certainly familiar with the general social logic involved. We all are, for we are all farmers of knowledge.

Sometimes, however, social conditions are such that the rain of suspicion develops into a sudden downpour, a flash flood, that threatens to sweep away both crop and soil—both self and knowledge. Sometimes conditions are such that the relationship between interests and knowledge becomes an unavoidable focus of one's attention, eroding the field of one's knowledge cultivation right down to the roots. PFI farmers report that this erosion is particularly intense in moments of economic stress and conflict, when the threat of others' interests to one's own seems especially salient.

It is not an easy matter to take up a different cultivation of knowledge. Most PFI farmers who have done so describe the period of change to sustainable agriculture as a great challenge, a kind of personal transplanting to a different field of knowledge cultivation. Many also describe the change to this new cultivation as a rapid breakthrough—as a sudden *phenomenological rupture* and subsequent sense of a new calling—with everything seemingly happening at once. It is difficult to give up an old identity and yet retain an old phenomenology of farming, and vice versa, because of the cultivation of identity in the cultivation of knowledge. So the two often change together. New farm, new self; new self, new farm.

But what is this new cultivation? What I argue in the last three chapters of the book is that sustainable agriculture is more than an alternative set of farming methods—ridge tilling, organics, rotational grazing, etc.—and an alternative set of social relations, of friends and associates with whom one identifies. It is both of these, but also, and perhaps more important, sustainable agriculture is a different *social practice* of agriculture—at least as practiced by Practical Farmers of Iowa. By a different social practice, I mean that the relations of knowledge within PFI have a different feel to them, a different way of experiencing others and of experiencing one's own self. And that different way is to recognize difference and to encourage it as a source of learning, change, and vitality, rather than as a threat to self and knowledge. That different way is the way of *dialogue*, rather than monologue.

In an uncertain world, however, monologue can be very appealing. Doubt is a difficult terrain on which to maintain a sense of self. Industrial agriculture offers listeners the monologic comfort of its universal claim to truth, secured on high in the laboratory and in the market.

Many also find monologic comfort in accepting only one's own word; at times, we all do. A focused recognition of the interests behind knowledge can bring about a loss of confidence in the words and experiences of others. Or it can create a self so uncertain that it finds stability in the denial of others—in a retreat into one's own self and one's own farm, abandoning science, government, industry, even one's family and friends for the pure local knowledge of the farmer himself or herself.

Either version of monologic practice—the one universalistic, the other solipsistic—leads to a similar social condition: that of disengagement from others and their differences. They lead to speaking without listening, and listening without speaking. In either case, we are left with nothing to talk about.

There is a third possibility, though, and most PFI members are discovering it. The recognition of interests can enrich the soil, as it were, by encouraging perspectives that blend their knowledges together. Rather than rejecting the social relationship between interests and knowledge, as the universalist does, rather than rejecting the possibility of knowledge because of its social relations and interests, as the solipsist does, and rather than rejecting the social itself, as both these forms of monologue do, most PFI farmers seek dialogue.

Central to this new understanding is how PFI farmers understand themselves as men and as women. The phenomenological investments of the self are strongly gendered in rural Iowa, as in most places. Those gendered investments typically encourage monologue between and among men and women. They encourage men to find their selves in the seductions of authority, of speaking without listening, of asserting sameness across difference, of having little need or concern for attending to the needs and concerns of others. They encourage women to find their selves through supporting the monologues of men by serving, paradoxically perhaps, as both audience and stage manager for them. But PFI farmers, men and women, increasingly find this an unsatisfactory arrangement for all parties, as it suppresses the vitality of difference that they are discovering, through dialogue, is very much worth relishing.

PFI farmers seek therefore to create a *dialogic agriculture*, an agriculture that engages others—men, women, family members, other farmers, university researchers, government officials, and consumers alike—in a common conversation about what it might look like. The aim of PFI farmers is thus not to create the top-down universal truth of the absolutist, nor the exclusively bottom-up local truth of the solipsist. Instead, they seek to create the social conditions of an agricultural knowledge that endeavors to take into consideration everyone's experiences and everyone's interests, creating a cultivation of cultivations. Through dialogue, these PFI farmers seek to cross-breed knowledge, to create pragmatic knowledge that gets the crops to grow in ways that sustain families, communities, societies, economies, and environments.

In short, PFI farmers are great talkers, and thus better doers. Herein lies the heart of the cultivation of *practical agriculture*, an agriculture that roots action in dialogue and dialogue in action, and thereby sustains them both. In an increasingly fractured and untrusting world, this is a cultivation worthy of the interest of us all.

A few words about methodology. In keeping, I believe, with the spirit of PFI, this book is a species of what has come to be called "participatory research"—that is, research in which the people under study help conduct the study.[21] Two of my colleagues in this project, Donna and Sue, are longtime members of PFI. Donna is a former member of the board of PFI and farms in western Iowa with her husband and son. Sue has been an active volunteer in PFI since its beginning, and her husband, Rick Exner, was one of the people involved in founding the organization in 1985 and was long PFI's only employee. Greg and I represent more the external and academic side of the project, Greg as a former graduate student in sociology at Iowa State University and now an assistant professor of sociology at the University of Wisconsin–Fox Valley, and I as an associate professor of sociology at Iowa State, and now at the University of Wisconsin–Madison, where I am an associate professor of rural sociology. Neither Greg nor I had any previous personal connections to PFI or farming before the research began.

I should emphasize the phrase "represent more": all four of us have academic degrees in sociology and have close connections with Iowa State University. Donna has a bachelor's degree in sociology, Greg now has a doctorate in sociology, and Sue has a doctorate in rural sociology, all from Iowa State. I taught sociology at Iowa State for nine years before moving to the University of Wisconsin–Madison and served as Greg's and Sue's major professor. Plus, to study PFI and sustainable agriculture is to study the university itself, given the longtime role of land grant universities in agricultural research and the special association of PFI with Iowa State University. And as the research progressed, my own personal ties to PFI increased when my wife took a position at Iowa State in sustainable agriculture education, a position that led her to work closely with PFI on a number of projects. I also joined a folk band with Rick Exner and two PFI farmers, and we performed at several PFI events. For Greg too, his interest in sustainable agriculture and his personal participation in it increased over the course of the research. Although we did not initially conceive of the study in this way, it turned out in the end that we were all participants of one sort and degree or another.

Participation is not the usual way in which research finds its results, of course. It has long been a reflex of science to isolate its subjects—to turn them into objects neatly detached from the observer, supposedly fostering a neutral, value-free perspective. As I discussed earlier, such a conception of the actual practice of science is at best inaccurate. Science is a human and therefore social endeavor, with all the unavoidable subjectivity that goes along with being human and social. But even more, such enforced distancing can also foster misunderstanding and irrelevance. The allegedly detached observer—who in many ways is better understood simply as someone who is new to a social setting—may miss much that a person local to the scene would readily be able to point out. And conversely. The most apt insights should come from maintaining distance and closeness at the same time, drawing on the strengths of both.[22] That balance is precisely what participatory research seeks to attain, involving those with varying degrees of distance and closeness in the research through a process of open and critical interchange. As well, participation promotes research that is more rooted in the concerns of the community under study, and thus is more likely to be of direct social relevance—more likely to be research that is valu-able, and not merely value-free.

There is no precise formula, though, for how to conduct participatory research. It depends on the opportunities and constraints of each situation—on who has the time, the support, and the inclination to participate. But the intent is always the same: to democratize the research process through dialogue. The emphasis on dialogue does not mean, however, that all the participants have to agree. On the contrary, the point is to welcome discussion about differences of experience and opinion so as to better develop the critical sensitivity that is the true goal of all scientific research.

Nor does participatory research mean that everyone has to have the same role in the research process. At least it did not mean it here. I was "principal investigator," as funders put it, on the grant that supported our research. I initiated it, I was most centrally responsible and accountable for its conduct, and I did the writing. And why? Because part of my job as a professor is to conduct research and write books. (Also, I like to do these things.) Donna, Greg, and Sue were paid out of our grant—Sue helped write the grant proposal, too—and that income supported their efforts in helping conduct the field work and in participating in our many hours of meetings about what it all meant. They also volunteered their time in reviewing the drafts of the manuscript as the chapters slowly emerged from my computer. But the only financial support for the actual writing of the book was my salary as a professor. Besides, I've done this kind of thing before. So, for better or for worse, out of this tangle of positions and responsibilities and interests and experiences, the writing fell to me.

But the dialogue still isn't wide enough. One of the goals of this book is to provide an opportunity for others to engage in the conversation about what a practical, dialogic agriculture might look like—to broaden our understanding of who is a farmer. Iowa is far away from most of America, culturally and physically, as are most of the farming regions of our country. We read about Iowa and farming life in *The Bridges of Madison County, A Thousand Acres,* and *Moo.* We see them in *Twister, Field of Dreams, State Fair,* and a flurry of news items every four years about the Iowa caucuses. But few have actually been to Iowa and to farm country, aside from some dull hours on Interstate 80 between Chicago and the Rockies.

And yet it is there with us three times every day (or more, if you're like me) when we bring the rural deep into our urban mouths, although we may scarcely think of it. Food connects us all to agriculture. What we choose to eat and not to eat has enormous implications for what goes on in places like Iowa. Food makes farmers of us all, whether we are aware of it or not.

As an Iowa native once remarked, "The problem, then, is how to bring about a striving for harmony with land among a people many of whom have forgotten there is any such thing as land, among whom education and culture have become almost synonymous with landlessness."[23] That Iowa native was Aldo Leopold, writing some fifty years ago, and his words still ring true despite the ensuing decades of environmental debate. Leopold's point was not, I think, that we urbanites all need to go live part-time in an old shack in the countryside, as he did, or as Thoreau did a century before him. Rather, he was writing about the need for recognizing that none of us is apart from the land, however high above the ground our apartment may be or however far a trip it may be to where we can see crops grow. Moreover, none of us is apart from those who grow those crops, although the patterns of our lives may mean that we rarely encounter them.

Not apart, yet still very distant. This is the trouble with so much of contemporary life. We specialize. We categorize. We segment markets. We change channels by remote control. We gate our communities. And we meet each other in traffic jams, horns blaring. But what are we to do? It really is a long way to Iowa now, and to most of the other states of being in our society.

In other words, social life is a structured life and dialogue has to take place within its structures—even when dialogue's goal (as is often the case) is to change the structures of our lives. It's a paradox, and it often as not means that change doesn't happen. But sometimes it does. We can only hope sometimes it does. For sustainability is not about the maintenance of the past, but rather about the maintenance of the future.

Intermezzo

A book is both a product of, and a contribution to, the continuing history of human conversation. Some of that conversation is written down, as in the case of a book. In this sense, a book is just the writing down of conversation. But most conversations never get written down. One way to understand the goal of academic research is as the effort to record that which is not recorded, for if human wisdom is the accumulated result of conversation (as I think it must be), we are that much the wiser the more of us we can keep at least potentially active in the conversation of social life. Human memory being what it is, and the physical problems of bringing together all speakers, present and past, being what they are, *writing it down* greatly increases that active potential.[1]

But even a book, long as it may be, has its conversational limits. Indeed, that length is one of its limits, as not everyone has equal time and enthusiasms for a book-length contribution to the conversation on a given topic. Those differences in time and enthusiasms will lead us all to respond differently to a contribution to that conversation. Which is just great, as great as anything could be, for it is our differences that make conversation worthwhile to begin with. But it does present something of a barrier to the printed conversational contribution of a book, which cannot shape and fit its words to the time and enthusiasms of each individual reader.

With that barrier in mind, in this intermezzo section and three scattered later in the book, I will address some issues that I judge many readers will find less central to their enthusiasms, even as some readers will find these the most central to theirs. In the main, I will engage here the more academic concerns of theory and methods. Readers who are not enthusiastic about such matters can safely skip these four intermezzos, as they indeed may have already skipped the second half of the Overture. But I hope they may have a moment to let their eyes linger here, at least briefly, and thus contribute to the active potential both of these ideas and, I believe, of the human conversation on sustainability and agriculture.

Although I don't mention it in the Overture, my argument about dialogue and sustainability is in part directed toward the great gulf between modernism and postmodernism. This point may seem quite removed from the lives of Iowa farmers, and in a sense it surely is, for these are not the terms by which they consider their circumstances. But in another sense it is not, for the terms *modern* and *postmodern* are central protagonists in the debate over what to make of our confused epoch, and Iowa farmers must contend with this confusion like everyone else. At any rate, I found this debate unavoidable as I considered what to make of the current conditions of agriculture and sustainability.

The modernism-postmodernism debate, it must be said, is one that many have found tiresome—even some of those who are intrigued by matters of theory—and with some justice, given its polarizing tendencies. As I considered the debate, I too at points found myself slipping into the common oppositions of those tendencies. I believe, now, however, that I have come to a perspective that brings these polarities, if not into unity, at least into some connection.

For example, on the one hand, there is much about my argument which rings (some might say smacks) of postmodernism. Consider some of my basic points. That knowledge is social. That there is no universal perspective on knowledge, such as modernism has long claimed. That knowledge is related to, even relative to, interests. All of these points combine in Michel Foucault's observation about what he termed "power-knowledge": "There is no power relation without the correlative constitution of a field of knowledge, nor any knowledge that does not presuppose and constitute at the same time power relations."[2]

But that's not all I'm saying. I'm also saying that both modernism and postmodernism can lead to the ending of conversation, to monologue. Neither the assault of universalism nor the retreat of solipsism is the route to engagement, to dialogue. And there is much of the universalistic assault in modernism, given its commonly rough-shod ways, and there is much of the solipsistic retreat in postmodernism, given its commonly isolationist ways.

What does this have to do with agriculture? The connection of universalistic modernism to agriculture seems plain enough. Industrial agriculture's big markets, big technology, big science, and get-big-or-get-out farms are all moved by the all-encompassing, all-inclusive, all-consuming borderless logic of the universal—of truths that supposedly hold everywhere and of ways of being and doing that attempt, so it seems, to gain place everywhere. But the connection of agriculture to postmodern solipsism may be less readily apparent, mainly because of the success of industrial agriculture's modernist dominance. I have in mind here the supporters of a pure "local knowledge" agriculture based only on farmers' own knowledge, and aimed at "putting the first last," as one author has put it.[3] These supporters argue for the importance of creating "an alternative knowledge system that functions primarily outside of the dominant institutions of agricultural research and extension," in the words of Neva Hassanein, one such advocate.[4] Instead of the top-down knowledge of the university, the state, and the corporation, their goal is to create local networks based on farmers' bottom-up knowledge.[5] From such a position, engagement with top-down knowledge is a moral and intellectual hazard. Thus Hassanein worries that dialogue between farmers' local knowledge networks and those outside "can potentially undermine the autonomy of the networks and the direction their knowledge development takes," and that these networks "will require a deliberate effort to maintain and defend the autonomy of the spaces in which they exchange knowledge and build community."[6]

My point here is not to reject the value of local knowledge—far from it. But neither is it to reject the value of agricultural research and extension. Rather, it is to question the categorical and the absolutist and to invite discussion, disputation, and the active potential of dialogue. It is tempting to put industrial agriculture (and modernism) on one side of some line, and local knowledge (and postmodernism) on the other side. But this is a temptation we should resist or, where it already exists, seek to overcome.

Nor is the point to escape from the high-walled castle of the categorical only to land in the muddy moat of undifferentiated holism. Thou shalt not commit a category: this seems to be the unhelpful advice of much postmodernism.[7] Every category is in the end just an assertion of, or product of, cultural power, these writers imply. But there is nothing wrong with difference. We need difference, if we are to have dialogue. Difference gives us something to talk about and someone to talk with. Difference gives us the ends and means of what is said and what is thought. Without difference, we cannot think and we cannot talk. But we need difference without it becoming oppositional, so that those ends and means retain their active potential for change and for questioning the absolute. And when we engage in the universalism of unrestrained industrialism or the solipsism of pure local knowledge, difference is cast in such oppositional terms that dialogue ends.

In any case this is the vision of sustainable agriculture that PFI, at least in its better moments, tries to put into practice. Rather than valuing only the top-down knowledge of universalistic modernism or only the bottom-up knowledge of solipsistic postmodernism, PFI's goal is to turn the whole business sideways and get it all on the same plane, where people can talk things out in a relatively non-oppositional way.

Does such a non-oppositional approach mean the end of conflict? Not at all. PFI's approach neither prevents conflict nor is afraid of it. Such an approach recognizes that, quite often, difference is part and parcel of conflict. And that conflict is not necessarily a bad thing. We can learn from the difference that conflict creates and the conflict that difference creates—but not if we are so oppositional that we can't talk with each other about it. Difference can be practical, if we handle it right.

Part of handling difference right is not exaggerating it. In that spirit, let me take a moment to emphasize that not all postmodernism is solipsistic and that not all modernism is universalistic, or at least that these are not necessary implications of the modern and the postmodern. In fact, for many writers the whole point of postmodernism is to engage others and their points of view, and to learn from them.[8] This perspective might be termed *practical postmodernism,* "postmodernism you can do something with," as my colleague Michael Gardiner and I put it a few years ago in a book about the Russian philosopher Mikhail Bakhtin, from whom I derive many of the ideas in this current book.[9] Indeed, I long toyed with describing the position of this book with this term, seeing it as, well, postmodernism Midwestern style, merging postmodern cultural concerns with the Midwest's long tradition of pragmatism.

Yet another of the main goals of this book is to point out the intimate relationship between knowledge and identity, so central to the cultivation of knowledge, and how this relationship can isolate us in the categorical and the absolute. Cultivation is inescapable, but our work on this book has led my colleagues and me to value forms of cultivation that are maximally open to other identities and knowledges—forms that are as dialogical as we can get them. And it seemed to me to be, in the end, just a statement of identity to choose to call PFI's approach "practical postmodernism" instead of, say, *dialogic modernism*—modernism you can talk to. I saw no other basis for choosing between the two, as they both reach for the same dialogic position, and try to get there from different sides in the same theoretical conversation.

An influential group of theorists—Ulrich Beck, Anthony Giddens, and Scott Lash—reached a similar conclusion a few years ago, but in the end they did in fact choose a side, what they termed *reflexive modernization.* By "reflexive" they meant a modernism that reflects on where it's going, a modernism that is self-conscious and self-critical.[10] This indeed sounds like a modernism you can talk to, and I admire much about their approach. Yet I find that, while similar in its aims, their decision to include the word "modernization" in their term is indicative of several differences with the position that we, and we believe the people of PFI, are trying to get to.

The most important difference is that Beck, Giddens, and Lash explicitly intended to distance themselves from postmodernism, in part because they were frustrated by the solipsism they saw in it.[11] So they meant the term as deliberately oppositional to postmodernism. They also felt that it was not empirically accurate to say that Western societies are in some sort of period that is *post*modern.

There still seems to be a heck of a lot of the modern running through the current day, they argue, and of an increasingly reflexive sort, in that Western societies no longer have the same unquestioning confidence in the top-down expert knowledge of old-fashioned universalistic modernism.

While I agree empirically that Western societies have by no means left the modern behind, it seems to me equally inaccurate to proclaim that Western societies now routinely question and debate science, technology, and economy and the conflicts they create in some kind of open and self-critical way.[12] Such self-critical awareness is not yet part of the reflexes of the day, but rather, when it happens, represents a major struggle against the continuing power of the modernist monologue. Reflexive modernization may be something to hope for, but it isn't here yet.[13]

Not that my colleagues and I object to a bit of normative optimism—not at all. We seek that too in the terms we use in this book. But we may have a difference in the norms of our optimism. At least to me, the "-ation" in the term "reflexive modernization" seems to retain an echo of modernist directionality, of the modernist promise that it knows what is good for everyone else. Whether intended or not, the term reflexive modernization feels like it still has a kind of faith that everyone wants us—"us" being the industrial capitalism of the West—to sell them a two-car garage and a cell phone, but now we're going to stop to ask their permission first. In other words, it feels like it still seeks the universal and the absolute, albeit with a bit more realism about politics.

And I also feel that the very reflexiveness that "reflexive modernization" seeks is undermined by its explicit identification with one side in the modernism-postmodernism debate.[14] For this reason more than any other I have decided to drop the word modernism entirely, in both its original and "post-" forms. We still need some words to describe what we are talking about, though. Categories always have their oppositions and identifications, yet we do need to talk. We do need to decide on some words to say if we are to say anything at all. But let us choose words that contain within them an invitation to response and to categorical change. My decision, then, is to call what we're talking about the only thing that people ever talk about: *dialogue* and its *practical* implications.

PART I

THE UNCERTAIN LANDSCAPE OF INDUSTRIAL AGRICULTURE

1. Economy and Security

Somewhere in *Blue Highways,* his classic of American travel, William Least Heat Moon remarks that one can make a fair guess about the quality of small-town restaurants by the number of local calendars on the walls—the calendars put out by the local chamber of commerce, the local little league, the local school district, the local Farm Bureau chapter, the local car wash. Local people know where the good food is. If there are lots of local calendars on display, it's a place where the local people go, Moon figures, which means you're more likely to find fresh coffee, French fries made from actual potatoes instead of reconstituted potato powder, and apple pie with a hand-rolled crust. Moon suggests assigning one star for each calendar.

By this measure, the J and D Café in Powell Center, Iowa, is a four-star establishment. The Norman Rockwell image pretty much stops there, though. Powell Center is the county seat of Powell County, a struggling town in a struggling county in western Iowa. It's holding on enough that at least four local organizations and businesses put out calendars of some sort, but plywood is one of the few items that still sell briskly. There are two boarded-up stores on the right of the J and D, and one on the left. The J and D isn't exactly thriving either. If the number of different kinds of wall covering were another way to rank local restaurants in small-town America, the J and D would be a five-star place. There are three kinds of mismatched wood paneling, a section of painted plaster board, and a patch of wall covered in orange shag carpet—products of a kind of continuous remodeling project by Jake and Deanna, the J and D's proprietors.[1]

But by any measure, the J and D was a lively place to be on the March night that I was there with a group of Powell County farmers and farm couples. A friend from Powell had heard I was doing research for a book about what it's like to be farming at the turn of the new century in Iowa, as I often described this project, and she invited me down to play volleyball with some local farmers who get together in the Powell High gym most Wednesday nights during the winter. "It will be a good way for you to get to know some of our farmers informally. That way they'll open to you more," she advised sensibly. Besides, I like to play volleyball and they were short of players. We played for a hour and a half before heading, sweaty and happy, for the J and D. Jake and Deanna had set up a long table in the middle of the restaurant for the group of about eighteen of us. I took a seat somewhere in the middle. The beer was cold and good.

"So, what's this book about?" someone asked.

I made some joke about not knowing, as I hadn't written it yet, and then gave my standard line about trying to get farmers' stories of what it's like to be farming at the turn of the new century.

"Well, there's not many of us left with stories to tell," said Clint, the grain farmer on my right, a man in his mid-thirties. Several others laughed.

I had heard this kind of gallows humor before among farmers, so I followed up with, "That's why I'm doing it—interview you before you're *all* gone!" I immediately froze a bit, afraid I'd gone too far, as I heard myself talking faster than my mental editor could keep up with after half a beer. But everyone laughed appropriately. I relaxed, and they seemed to as well.

The conversation near me soon took off from the idea of farmers all going out of business to the threat posed by the coming of large-scale hog confinement operations, or what the media call "hog lots." Iowa leads the United States in pork production, and for many years raising hogs has been one of the main ways Iowa's farmers have sought to do something more profitable with their grain than selling it to the local elevator as a raw commodity. As recently as the early 1990s, there were some 35,000 hog farmers in Iowa, about a third of the farms in Iowa at that time. By 2001 the figure had dropped to about 10,500, as the industry has taken to raising hogs like chickens.[2] Most Iowa hogs are now reared in massive metal barns with thousands of animals living at close quarters on a slat-metal floor, never seeing the light of day. "Slat barns" are what they're called in the trade. The feed pours in from bins standing above the barns, and the manure pours through the slats into pits down below. From there, the manure is usually funneled outside into waste "lagoons." Inside these lagoons, the manure often takes on an anaerobic stench that can be smelled for miles. In the past ten years, this stench and the economic concentration in agriculture that it represents has become one of the biggest political issues in Iowa.[3]

"You see what's happening," Clint observed to Eldon, a local plumber who used to be a farmer, sitting across from him. "Pretty soon they'll control the whole market, and that won't be good for anybody, in my opinion. They'll be able to charge what they want."

Eldon nodded but didn't reply, and Clint took up another conversation with the farmer on his other side.

Then Eldon caught my attention and said to me, "I can understand why somebody in a city, like Ames, doesn't like it," referring to the big hog confinements.

Surplus corn piles up outside the Pro Co-op in Rolfe, Iowa, 2001.

Ames is where I lived at the time. It's also the location of Iowa State University, with its College of Agriculture, then my employer. There had been a couple of recent controversies over the construction of hog lots near Ames. Eldon knew all of that, I think, and probably chose his example accordingly.

"Because they didn't expect it," he continued, checking to make sure that Clint wasn't listening. "But in the country, well, it's part of life here. And the jobs depend on it." Eldon searched for words for a moment, and then said simply, "It's progress."

It was my turn to nod, hoping my surprise wouldn't show. Here was a former farmer advocating economic concentration in agriculture, saying it created jobs. But economic concentration was probably one of the main factors that had pushed him out of farming. I wondered if maybe the big hog confinement barns required a lot of plumbers to set them up, which might explain Eldon's apparent acceptance of them, and I considered how to inquire delicately about that.

But while I did, there was suddenly one of those group pauses in a room full of conversation—a happenstance moment when nearly everyone at the same time stops to think of what to say next, and everything gets quiet. As sometimes happens, one voice didn't notice the unexpected break in the hubbub. Clint's excited, beery voice came charging through to all of us.

"I tell you, I have half a mind to try to outbid that wimp for all his land!"

I have no idea whom Clint was talking about, and I don't know if anyone other than the farmer next to him did either. But it was plainly a pretty shocking statement to everyone around me. Most Iowa farmers rent a lot of the land they farm, if not all of it. To outbid a farmer for rented land is to threaten a livelihood. Clint, who already farms well over a thousand acres and is thus well established, was probably in a good position to offer a few extra dollars' rent and drive most anyone he wanted to out of business. But you don't say these things openly. Eyes met briefly up and down the table, and then looked down, around, anywhere.

Clint seemed to realize that he had crossed a moral line. He shrugged and turned to me as a sort of neutral figure and said, "Farming is cutthroat in this county."

That was something everyone could agree about, apparently, and the conversation took off again. Several people chimed in with stories of land rents going up and up and up, mainly directing the conversation my way. Instructing me about the competitive conditions in the Iowa countryside seemed to bring the group back together. Eighty dollars an acre, they told me, was the current average price in the county, with some land fetching up to double that in recent months. (At this writing, the average rent is up to $140 an acre, with some good ground fetching more than $200.)

"If I understand you right," I said after a few minutes of this, feeling provocative, "these days the farm next door is your competitor, not your neighbor."

Again I worried that I'd gone too far, and again I need not have. "Your competitor, definitely," said someone in reply, and several others joined him in a laugh that was equal parts uncomfortable and ghoulish.

"We just keep bidding each other up for land," explained Walt, a quiet-spoken farmer with a midsized farm who was sitting to my left.

I don't know why it popped into my mind, but I decided to ask, "Well, if you're driving each other out of business like this, why don't you get everyone organized and refuse to pay such high rents? Why don't you decide to pay, say, eighty dollars an acre and no more?"

It was a naïve thing to say, and people looked a bit startled for a moment. I probably had gone too far with that one. But finally Walt offered an answer, using the generic pronoun of choice in most farming circles, "Well, then he'll come in at eighty-two dollars, or whatever, and price me out of it."

There are some things that enjoying a beer after volleyball with your friends can't change—or at least that evening hadn't yet.

The things that playing volleyball with your friends can't immediately change are forces larger than a game, larger than a community, larger than a state, larger even than a nation. Iowa farmers face the same basic issues of contemporary American life that we all do: downsizing, escalating economic competition, the pervasive sense of never having enough time for anything, conflicts in the relationship of home to work, struggles to accommodate difference, loss of community, and resulting threats to self and identity. Farmers are us. The new uncertainties of life in a contemporary capitalist country are theirs as much as anyone's.[4]

But farmers face these issues in ways that are peculiar to their work and to the rural scene. Farmers are not us even as they are us. The main effort of this chapter and the two that follow is to trace these new uncertainties as they are expressed in the lives of the sixty-nine Iowa farm households my colleagues and I studied for this book. Underlying these uncertainties are the market, government, and technological forces that rural sociologists and agricultural economists like to call the "structure of agriculture." There is more to these uncertainties than matters of agricultural structure, as I will describe in the second part of the book. But the effects of structure on the economy and culture, community and environment, and home and family of an Iowa farm are central to understanding why some farmers have switched to sustainable agriculture and why most have not, despite these challenges of uncertainty—and sometimes worse.

It is perhaps surprising to think of farming in the context of uncertainty, given the pervasive image of farming as a kind of respite from the churning movement of city life. The countryside is supposed to be stable, a place beyond the restless turmoil of the street, the highway, the ever-sprawling suburbs, the stock exchange, the ambitions of strangers. Rural life is supposed to be an eternal life, closer to the simple and changeless truths of nature. We have long believed that people are more authentic here—more natural for their closeness to nature—and that so too is the community, woven through with generations of social ties, families working together as they work the land. The country is home, we fantasize, our real home, our natural home. It is what we go back to, and it is supposed to be still there when we return, even if, as the historian Bill Malone has observed, we know it really isn't.[5]

Consequently, farming and rural life have a special status in our cultural imagination. Children's literature, for example, vastly overrepresents rural images, given the urban lives of most contemporary American children. The rural life we show our children through books like C. S. Lewis's *The Chronicles of Narnia,* Laura Ingalls Wilder's *Little House on the Prairie,* and Margaret Wise Brown's *Big Red Barn*—to mention three favorites of my own children—is alive, honest, unpolluted, nonindustrial, playful, natural, even magical, and, perhaps most important, safe for kids. Indeed, children seem to be themselves the cultural embodiment of this imagined world. Rural life is, at least in myth, *Babe*—a green and pleasant children's movie, where the animals can talk and the farmer is too kind-hearted to eat them.

This special status also emerges in the bewildering array of governmental and nongovernmental organizations devoted to promoting farming. I work for one of them, a land grant university, part of the system of institutions of higher learning set up by the federal government, beginning with the Morrill Act of 1862, to promote "agriculture and mechanic arts." (The Morrill Act gave each state at the time a grant of thirty thousand acres of land to sell or to manage in order to fund such a university; hence the name "land grant.") Every state and territory now has at least one land grant university, even the states and territories that weren't in existence in 1862: Iowa State University, University of Wisconsin–Madison, Ohio State University, Tennessee State University, Texas A&M, University of Puerto Rico, and University of Guam, to name a few.[6] Almost all land grant universities have now adopted broad liberal arts programs, but this is where you'll find a state's college of agriculture, if it has one, and most do.[7] This is also where you'll usually find most of the state's "extension" program, which every state does still have, and most of these retain at least some of the rural and agricultural outreach focus that was extension's original brief.[8] Then there is the vast apparatus of the United States Department of Agriculture, the departments of agriculture in almost every state, and the national agricultural census, all with their dozens, even hundreds, of sub-entities. And then there are all the national farm organizations—Farm Bureau, National Farmer's Union, National Farmer's Organization, 4H, National Corn Grower's Association, National Pork Producers Association, National Cattlemen's Association, and more, all with systems of local and regional chapters.

Of course, many of these are economic organizations, at least in part, and there can be no doubt that agriculture remains central to the U.S. economy, and to the human economy.[9] But from a purely economic point of view, there is something outsized about all the attention paid to farmers and to farming. Consider the absurdity of having a United States Department of Construction Work. And there are three times more construction workers in the United States than there are farmers.[10] Indeed, there are only some 2 million farmers in the United States, and about 1.2 million hired farm workers and agricultural service workers—all told, a tiny fraction of the country's 141-million-strong labor force.[11]

Or even consider an apparently purely fiscal matter like the U.S. Internal Revenue Service's distinction between farm income and nonfarm income, which gives every U.S. taxpayer at least a momentary annual pause as they contend with line 18—"Farm income or (loss). Attach Schedule F"—on their 1040 form. Behind the door labeled "Schedule F" is a vast labyrinth of special tax credits and deductions for farmers, albeit with accompanying special forms and bookkeeping requirements. True, farming has some special characteristics that do not fit easily in the square boxes of the tax code. But I imagine few professions feel themselves an easy fit with the tax code. And yet there is no line on the 1040 that says "Construction Work income or (loss). Attach Schedule CW."

The special status of farming is nowhere stronger than in Iowa, and perhaps with more economic reason than in most other places. Iowa is the number-one state in area of harvested cropland and in the production of hogs, corn, and soybeans.[12] It's usually number two in net farm income, behind California.[13] And it's the number-three state in number of farms, in the market value of agricultural products sold, and in the percentage of its population living on farms.[14] Agriculture and agriculturally related activities account for 25.6 percent of Iowa's gross state product and 22.8 percent of the state's workforce, according to a 1997 study.[15]

All of which sounds very significant, and it is. But the only way to get the agricultural portion of Iowa's gross state product and workforce that high is by including things like meat packing, tractor manufacturing, the shopping that farmers do, and workers in agricultural organizations. Even professors in the College of Agriculture at Iowa State show up in these numbers. Farmers themselves are less than 5 percent of the state's workforce, and directly contribute only slightly more than that to Iowa's gross state product.[16] These figures are about double those for the nation as a whole, but even within the agricultural industry farmers rank as Iowa's second-least significant economic category. According to that 1997 study, only workers in agricultural organizations—that includes professors again—are a less economically significant part of the agricultural industry. Farming is no longer the same thing as agriculture. It hasn't been for decades, if it ever fully was, but the disjunction grows wider with each passing year.

Nevertheless, Iowa surrounds itself with images of farming, such as its current state slogan, "Fields of Opportunities," or the one from a few years back, "Iowa: A Place to Grow." It would be far more accurate to surround itself with images from the other 75 percent of its economy, or at the very least from the larger parts of its agricultural economy, like meat packing. But meat packing hardly has the romance of farming, and not surprisingly it isn't even vaguely referred to in the state's slogan. (However, during the statewide contest for a new slogan that resulted in "Fields of Opportunities," the media had a good time with some of the spoof entries, including "Iowa—Eat Pork or Die.") Farming is an ever-present feature of the state's news. Ads for tractors, seeds, and pesticides soak up a huge portion of the available slots on local sports broadcasts—about one in three, by my informal estimate. Most adults in the state seem aware at any given moment of how the farmers are generally doing. During the Iowa caucuses, the candidates dutifully intone Jeffersonian truths about the importance of the farm in the life of the nation. And during races for Iowa political offices, the candidates are sure to play up any personal farm connection they might have, however remote.

My point here is not to belittle the importance of farming. Far from it. We vitally need what farmers produce. And farmers do have large indirect effects on the economy. But so too do we need what construction workers produce and, if you're a meat eater, what meat packers produce. Construction workers and meat packers also have large indirect effects on the economy—not as large as farmers perhaps (I have seen no studies making this comparison) but certainly substantial.[17] There is an economic dimension of farming's special status in Iowa and in the nation at large, but the hold of farming on our imagination goes well beyond the economic. It is also a matter of the cultural associations we make with farming as a part of nature, an act of simplicity, of honesty, of certainty, however romanticized and empirically inadequate these associations may be. Recognizing its cultural values enlarges, not belittles, the importance of farming for us all.

The special cultural status of farming is something that many farmers themselves agree with their city cousins about. I got to talking about this one morning with Wendell and Terri, a young Powell County farm couple. As we sat in the dining room of their white bungalow, the younger of their two boys, Evan, careened about, occasionally bombarding me with toys. (Wendell and Terri use their dining room as the children's play room, which compounded the intrusion I must have represented to Evan.) Their other boy, Tavin, was across the road playing at a neighbor's. I had asked them what they valued most about farming. Terri replied to my question with a laugh.

"Don't have to live in a city! I hated living in—"

"You lived in a city for a while?" I interrupted, surprised.

"For a year, when I went to school," said Terri.

"What city was that?"

"Des Moines, West Des Moines," she answered. Des Moines is the only city in Iowa of any significant size by national standards, with a population of about half a million in its metropolitan area. That means about one in six Iowans lives there, or in its rapidly expanding suburban and exurban fringe; West Des Moines is part of this expanding fringe. Des Moines is, frankly, not a particularly nice city, with little street life or night life or architectural life, aside from the stunning state capitol and a few other buildings. The governor of Minnesota at the time, Arne Carlson, touched off a storm of interstate jingoism in 1997 by declaring that "Des Moines is dead." Bill Bryson, the popular travel writer and humorist, begins his book *The Lost Continent* with the lines, "I come from Des Moines. Somebody had to."[18] Little loved even by urban enthusiasts, Des Moines, not surprisingly, is a symbol for many rural Iowans of all that is wrong with city life.

"That's a city all right," I said, and we both laughed.

"And I did not like it," said Terri.

Wendell nodded in agreement, and explained why. "I guess just some of the things about farming," he began, and then stopped himself for a moment.

"It's being able to appreciate the night," he said finally. "I've gotten to where I like to appreciate the nature. . . . I like the simpler things in life. I like farming. I like the theory of getting up some day and you don't have to go to town. You don't have to go to town if you don't want to deal with people. You can say the heck with it. I'll just stay out here and not have to worry about it. I guess maybe it's just a different upbringing than the city and its rat race."

Evan, who had been watching me intently from his mother's lap for the last minute, suddenly leaped up and ran out of the room. Wendell pointed at him as he went, and continued with another benefit he finds in rural life. "Being able to let the kids roam."

"To open up the door and let them outside," Terri added.

". . . This one there, Evan, he's used to roaming around outside here," Wendell continued, ". . . with his own creativity. You definitely have to keep an eye on him, but it's nice to let him do a little exploring, and learning other things."

Farming today isn't the same as agriculture, but neither is it *Babe* or *Big Red Barn*. Terri and Wendell's kids can't run off on their own at home without an adult keeping an eye on them. The farm is full of dangerous equipment and, for a young child, dangerous animals. (Terri and Wendell raise cattle and hogs in addition to some five hundred acres of corn and soybeans.) Terri and Wendell were not trying to say otherwise. Rather, they were using the imagery they had cultural reason to suppose I might understand to talk about their identification with a manner of living that seems to them to be disappearing.

Clinton-Garfield Cemetery, with the elevators of Rolfe, Iowa, in the distance, 2000.

Elsewhere in our conversation, Wendell put it this way. "I kind of have the sense that some of the larger farmers have got themselves to where, I don't know. It seems they're all wrapped up in getting the work done. They've got so much to do, I don't think they really totally enjoy it. Maybe they do enjoy it. Maybe they get a thrill out of driving that big equipment. But you're still removed from [nature]."

It's getting to the point, argues the rural sociologist Paul Lasley, a longtime observer of the farming scene in Iowa, that even agriculture isn't agriculture. "Agriculture is increasingly just 'ag,'" says Lasley.[19] The shorthand "ag" is taking over the agricultural lexicon. The press and the experts love to write and talk about "ag industries," "ag products," "ag exports," "ag markets," "ag statistics," "ag engineers," "ag equipment," or to use parallel locutions like "agribusiness," "agri-economics," and Conagra, the name of an "ag" corporate giant. Lasley's point is that this shortening of the word is an unfortunately apt reflection of the state of agricultural affairs. The culture part of agriculture is going, going, and, say many, perhaps soon gone entirely.

Perhaps we should see these changes as just that: changes, not losses. Perhaps it is just romanticism to see them otherwise, for it is just as cultural to enjoy the Big Machine way of agriculture as it is to enjoy styles of farming that bring the farmer close to "nature." There is a culture of machinery too. Perhaps we are just seeing a new rural culture developing, and there is no ground independent of culture from which to weigh one against the other. Perhaps country ways are always experienced as in decline, given our association of country life with stability—what the British sociologist Raymond Williams aptly termed the "golden escalator" of romanticism.[20] Perhaps so. Indeed, probably so, as I have myself argued elsewhere.[21] But Iowa farmers experience it otherwise.

The changing structure of agriculture has a lot to do with their experience of a disappearing rural culture. My interview with Wendell and Terri that day kept coming back to work pressures and this foreboding sense of threat, of loss of connection. After about an hour and a half, Terri and I were both tiring of the interview and thinking about getting on to other things. Evan had long since tired of it, and crashed into me with the toy lawn mower he had been pushing all over the house. Terri looked at me apologetically. "Don't worry about it," I said, but I had gotten the message. It was time to wrap up. So, by way of conclusion, I asked one of the standard ending questions of any interviewer: "Is there any question that you think I should have asked that I didn't ask?"

"We've pretty well covered it," said Terri.

But Wendell was still raring to go. "I guess the biggest thing we haven't covered is the shrinking number of farmers," he said, and then went on to make a series of notable observations that I think are worth hearing at length. Even Evan seemed to realize that his father was saying something significant, and he listened quietly as Wendell spoke, settling once again into his mother's lap.

"It looks to me now that agriculture in the United States is taking a little bit different angle than what happened over in Russia," Wendell continued, meaning to come gently to a surprising argument: that in fact what is happening in the United States is much the same as what happened in Russia. "Russia was governmentally run agriculture. Now, in the United States, the trend where we're heading right at the present time is instead of governmentally owned, it's corporate. Large, large corporations. Instead of governmentally owned, it's going to be a few factions that control it. That failed over in Russia. It didn't work. It was so inefficient. People kind of lose their work ethics when you're dropped to where you're really just putting in time for somebody else."

I was struck by the parallel Wendell was drawing between corporate control in the United States and government control in Russia during the USSR years, and their effects on the work ethic. I wondered where Wendell would go with it. He seemed to wonder a bit, too—people often think aloud in interviews—and paused to think his words through.

"We've still got a fairly good work spirit right now," he continued finally. "But I don't know if agriculture will ever be something you can really classify as a nine-to-five job. You're going to have to have the will to put in that extra time. You just can't have somebody sitting out there in a tractor and all of a sudden 'oh, no, it's five o'clock, and I got another hour or two of field work to do, but I think we're just going to call it quits today.' You can't just drive away, or it's not going to work. I don't know if we're a long ways from having that happen or not, but I got a feeling that we're getting to where people consider farming as more a job instead of a lifestyle.

"The talk is about contractors getting into business where they're going to almost do the total farming operation. . . . It'll basically eliminate the farmer. But then you're also dealing with hired employees to operate the equipment. Maybe they can do it with them, but I don't like the trend."

"So you're saying that you think there's a place for government in agriculture?" I asked. Farmers are notorious for their suspicion of government, and I wanted to know if Wendell thought government could serve as a balance against corporate power and absentee landlords, a way to protect the independent farmer. His answer located his politics in the realm of strong populist libertarianism, both anticorporate and antigovernment. But then he went on to say that corporate power is growing, and then to make a shrewd observation about the economics of agriculture today.

"I don't know. It seems like anytime the government gets involved, it may start out at the right thing but it gets just way out of hand. . . . I don't think government should run agriculture and I sure as heck don't think corporations should either. But at the present time it looks like there's going to be more and more contract production. Basically trying to do it for X many dollars. I don't know. Maybe the reason why it's in this situation . . . is agriculture's one of the purest forms of capitalism, where they got us all out here competing with one another to produce something the cheapest. Really it is. It's who's going to do some work cheaper. You know, you and I talked about the high prices for renting land around here. Well, if somebody's paying those high figures, then they can do it cheaper than I can. It really is kind of the purest form. Basic capitalism. Competition where you're producing for your cheapest."

Wendell was talking about what sociologists call the *treadmill of production.*[22] We're all familiar with the idea that competition stimulates producers to produce more cheaply and to adopt new technology to help them do that. The theory of the treadmill of production points out that this process of "basic capitalism" tends to lead to continual problems with declining profits and overproduction. Say you're a producer in a market, any kind of market, and you decide that you want higher income for some reason—your kids are going off to college, the house needs a new roof, or just plain old-fashioned greed and the desire for higher status. You have four basic production options:

- You could decide simply to raise your prices, without changing your production. But you might not sell as much then—unless you can work out a little price fixing through collusion with other producers. And that's illegal.
- You could decide to produce more so you can sell more, and that may help you for a time. But the increase in supply caused by your greater production will eventually lead to a fall in prices, because a lot of other producers will probably try the same strategy if they see you doing well by it.
- You could decide to adopt new technology that allows you to produce your higher output more cheaply. That way a fall in prices due to greater supply will still give you more income, because your own costs per unit of output are lower. But again, this will work only for a time, as other producers are likely to try a similar approach if they see it working for you, producing loads more product and causing such a drop in prices that everybody's profit margin, including yours, is soon wiped out.
- The only other production option is to adopt technology that allows you to produce at the same level of output as before, but more cheaply, resulting in the same prices for the same output yet a better margin. Great idea, but somehow you have to pay for that new technology, which gives you an incentive to try to increase output to make up the difference. And that only leads to a fall in prices, particularly when others start copying you, promoting the whole cycle all over again.

Actually, you don't even have to want to increase your income to find yourself having to increase output and buy more technology. You might be happy with what you've got. But if somebody somewhere in the market makes a move, pretty soon you'll find yourself running to catch up, as the treadmill starts to rotate under your feet. You and others in the market will start trying to produce more, probably by investing in better technology in order to do so, leading to declining profits once again as your costs increase and the price of your product drops with the increase in supply.

In sum, production-oriented solutions to the problem of the treadmill of production typically work only in the short term, and they have the long-term effect of making the struggle to stay on the treadmill ever more intense. But still farmers try them—this is basic capitalism, after all—every day ratcheting up the speed of the scramble that farm families like Wendell and Terri's must somehow endure.

Of course, another approach is to work on the demand side of a market, perhaps through advertising or through mounting an international trade expedition. Iowa's current governor wants to make the state the "food capital of the world" and he has just this kind of strategy in mind. Similarly, the agricultural commodity groups and other mainline agricultural organizations—the National Corn Growers Association, Farm Bureau, and the like—routinely call for free trade and expanded markets. But there's a problem with food: People can eat only so much of it. Economists call this problem "inelastic demand," and it presents a serious dilemma to agriculture. True, there are lots of hungry people in the world. But most of them don't have the money to buy more food, which is one of the main reasons people go hungry to begin with. Those with money to buy more food typically are already well fed. Population growth may help somewhat, but most of this growth is among those who don't have the money to buy more food, especially the kind of rich people's food that Iowa mainly produces: meat and the grain to produce it. (Besides, most people look to agriculture to save us from the dangers of population growth, not depend upon population growth for agriculture's own survival.) So advertising and international trade can only help somewhat.

Treadmills of production exist in virtually all sectors of capitalist economies—even academia, with its constantly rising expectations for number of publications, thus devaluing each individual work. But the inelasticity of demand for food, combined with the regional specificity of most agricultural production and the regional crises that weather often brings, means that farmers experience their own treadmill in a particularly intense and local way.

We could call this intense local treadmill the *farmer's problem,* and it runs like this.[23] If things go well on the farm one season—good weather, good technology, good production—chances are it also went well for most other farmers who grow what you grow. They're on the same treadmill, in the same general climatic region very likely (or they wouldn't grow the same crops), and they probably pay attention to the same major information sources about what technologies to employ (and so do their bankers). So you'll have a crop to sell, but lousy prices because of high supply. If things go badly on the farm—dismal weather, technology inappropriate to the season's challenges, and thus poor production—chances are it also went badly for other farmers in your region and growing your crops. In a way

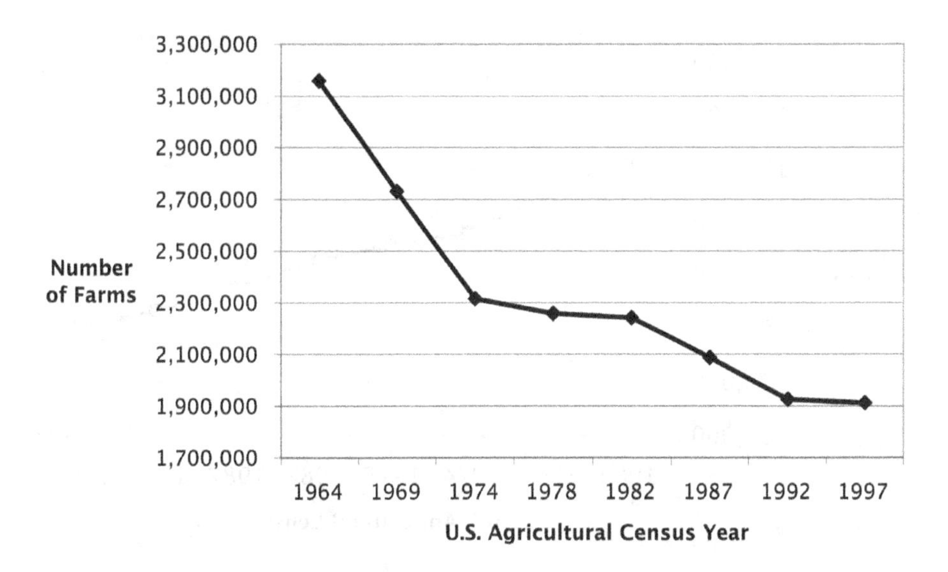

Figure 1. Number of U.S. Farms, 1964–1997

that's great because it means prices are high, as supply is low. But if you don't have much of a crop to sell (because it was a bad year) you can't really take advantage of the good prices. Consequently, the farmer's problem is that you pretty much only make money when you have a good year and a lot of other similar farmers don't. And that doesn't happen very often.

There is a solution, of sorts, to the farmer's problem: Knock a few producers off the treadmill, creating space for those remaining to grow bigger and maintain a decent income despite lower profit margins. This has long been the favored course of action in U.S. agriculture. Sometimes as a result of deliberate government policy—like the dairy herd buyout program of the 1980s—and sometimes as a result of more subterranean processes—like family conflicts, health problems, bad luck, bad management, and bad relations with the banker—the number of farms in the United States has been dropping steadily since 1910. The 1980s are often referred to as the period of "the farm crisis." But if one looks at the graph of the number of farms in the United States over time, this period shows up as only a minor blip on the line (see figure 1). True, the rate of decline in farm numbers increased during the farm crisis, as shown by the steeper slope in the line from 1982 to 1992. But this increase still left the rate of decline far below what it was from 1964 to 1974. From this point of view, the flatter slopes from 1974 to 1982 and again from 1992 to 1997 show up as uncharacteristically good times. And in Iowa, the state whose desperate farmers most often flashed across the nation's TV screens during the height of "the farm crisis," there was

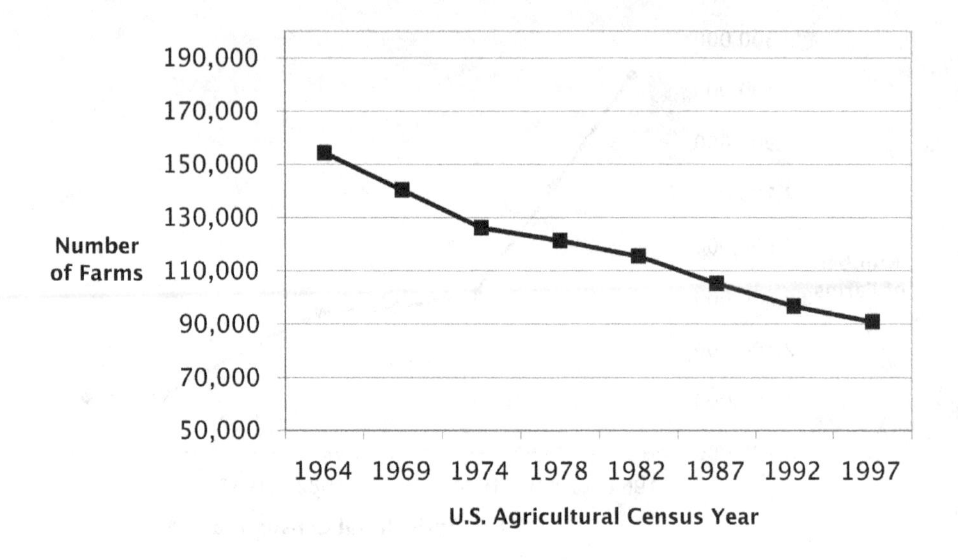

Figure 2. Number of Iowa Farms, 1964–1997

barely any increase at all in the rate of farm loss during the 1980s (see figure 2).

It would thus be more accurate to speak of a perpetual farm crisis. There was a mild uptick in farm numbers during the thirties, as the Great Depression sent some city folk back to the countryside for a few years, and there have been a few brief periods of somewhat better conditions, such as 1974 to 1982 and 1992 to 1997. But otherwise, U.S. farmers have been steadily eroding off the land, along with their soil, for the past hundred years. And every farmer still on the land knows it.[24]

Clint, the farmer who had had a bit too much beer that night in the J and D, certainly knows it. I later interviewed him back at his farm, sitting around the kitchen table of the suburban-style farmhouse where he lives with his wife, Rhonda, and their two kids, ages eleven and twelve. Clint is a lively and personable man, and we both seemed to enjoy the interview. (I know I did.) There was plenty of laughing and joking as we talked. He wanted to come up with his own pseudonym.

"I want something dramatic, like Clint," he said, striking a macho pose, "for Clint Eastwood."

"You got it," I told him.

But things were more serious as we discussed the competitive relations between farmers in Powell County.

"This farm is a real good farm where we're at now," Clint explained. "And I pay 135 dollars an acre for it. But it's got plenty," he stopped. "You're not giving out all my numbers, now?"

"I'm not going to give out numbers," I hurried to say, reminding him that I would be using pseudonyms and changing identifying characteristics in the book, including specific numbers, as I have done here. "But why is everybody so tight with their numbers?"

I was genuinely puzzled about this. Farming is now a commodity-based industry in which the profit margin depends upon the actions of millions of farmers across the country and even the world, in addition to the actions of corporations, consumers, and governments. What difference does it make to the farmer's problem, therefore, if the neighboring farm produces a bit more or less, or invests in some new technology or another? The actions of the neighbor are insignificant in comparison to the massiveness of the treadmill of Big Ag. Besides, everyone sells to the same few buyers at the same few prices set by the same few institutions, like the Chicago Board of Trade. There are occasionally special deals, but these are usually for specialty crops, like organic produce, aimed at niche markets, not the standard crops most farmers raise. So why should one farmer feel competitive toward the neighboring farmers? In other words, why is the farmer's problem local?

"Because my neighbor will bid 145 bucks for this son of a gun," Clint replied. "If he knows [what I pay], he'll bid ten dollars more. Ninety-nine percent of the time all I have to bid is 135, and my landlord'll let me have it. But if your neighbor is going to offer more, then that's what the landlord's going to want. And this farm would be worth that."

This was almost word for word what I had heard in the J and D. And talking to Clint, I suddenly understood. At least in grain farming, the competition is not primarily over selling one's crop but over the land base that makes it possible to have a crop to sell.[25]

In Iowa, slightly more than half of this vital land base is rented, in two main ways.[26] In the past, it was common for farmers to rent on a "crop-share" basis in which the farmer and the landowner split costs and profits, usually fifty-fifty but sometimes sixty-forty or forty-sixty. The trend now, however, is toward "cash-rent," in which the farmer pays a flat fee for the land and pays all the input costs, taking home any profit on top of that. Crop-share typically is based on a more personal relationship, with joint decision making and often long family and interfamily ties. But cash-rent, says Clint, "is a lot simpler." The decision making is easier. The bookkeeping is easier. And it's more businesslike. But it has its downsides.

"If you have a real good farm, the only thing bad about cash renting right now is that everybody's starting to be a cutthroat," Clint explained. "There's people running around bidding more. As to when you're sharecropping it, I mean that's what you do. You know where your acres are."

Those acres give you security—security of tenure but also security through the lower cash-flow demands of crop-share rental. Clint farms many more than the magical thousand acres, but he owns very little of it. Most of his land is cash-rent. This means he has to come up with nearly $200,000 every year for the rent alone, given that rents were running up to $150 an acre when we had this conversation and are considerably higher now. On top of that, there are the costs for equipment, seeds, fertilizer, and pesticides. (Clint is far from an organic farmer.) For example, the 280-acre piece where Clint's house sits he used to rent on a crop-share basis until a few years ago. Since it went cash-rent, the cash-flow demands on it have skyrocketed.

"For this 280 acres here, it roughly took me fifteen thousand to put my half in when I was farming it, you know, fifty-fifty"—in other words, when he was farming it as a crop-share rental. "But now when I have to pay half the cash rent, and then all the inputs of the 280, you're up to around fifty to fifty-five thousand, probably."

And that's only half the cash-rent. Clint is talking here about his springtime bill for inputs and cash-rent. He'll have to pay up the other half of the rent in the fall, plus any fees for contract harvesting on acres he doesn't have time to get to himself, and any other costs he may accrue in the interim (probably quite a few). And if it looks like he's making too much money—because the price of corn goes up, say—the landlords may set the rent higher for next year.

"With the sharecrop, you don't have to negotiate the rent every year," Clint explains. "Your landlord pays for half the input and you pay for half of the input. Of course, you do it with all your own machinery and stuff, but you don't have to worry about rent going up. You just know there's none of the negotiating around everything. Some cash-rent farms have a two-year lease, and some are three. But with the price of corn rising, the landlords are wanting short leases for next year. And if your lease is coming due, they'll want more cash rent. That's just the way it is."

In other words, Clint isn't just in competition with other local farmers. He's in competition with the local landowners.

And the landowners are in some competition with each other. It may not be economic competition, at least not much. As far as they're concerned, it's a seller's market for land. But in the view of Clint and most other farmers I spoke with about this, the landlords—a lot of whom are retired farmers themselves, and still living in the community—compete with each other for the status of having high rentals. Having high rents suggests that you're a shrewd businessperson, that your income is good, and that your land is good, too—all of which brings a measure of honor in the community. Most landlords my co-researchers and I spoke to about it also agreed that there is a lot of status competition among them, but here's Clint on the subject.

"I know of a guy up north a few miles—this is a tremendous farm—but he happened to pay 150 bucks an acre for cash rent. So now what do you think all the landlords are going to want? All these old retired guys will want 150 to go on renting. It's just a vicious circle."

Clint's main competition is with other farmers, however, hungry for the land they need to "stay in the game," as Iowa farmers say.

"People just are so competitive right here," Clint explained. "You know, I've always rented a farm. If it's for rent, you know, I'll rent it. Or I'll try to rent it. But a lot of people will drive around, like the people that own this farm," and he indicated with his hand the farm across the road, visible through one of the kitchen windows. "They'll call in and offer more money. I mean, there's people that will stab you in the back to get the farm. And that's," he shook his head, and finally said, "it's a bad deal."

"Wow," was all I could say in return.

"And, you know, that's life. But there is a lot of it."

Still, I couldn't forget Clint's own beery words in the J and D, and wondered to what extent his representation of the situation was in part a way to justify to me or to himself, or maybe to both of us, his own aggressive attitude toward land. I didn't know how to ask about it, but Clint brought it up himself a few minutes later, perhaps because he too remembered that night in the J and D.

"If there is another couple hundred acres of good ground, I, you know, I'm not going to go stab anybody in the back to get it," he said.

"Right."

"But if there is something close, and it's for rent, I'd take it. And that's how everybody else is here in the neighborhood."

Clint is not someone whom his friends and associates, I suspect, regard as a gentle, easygoing man. Nor do I suspect he sees himself that way. (His choice of pseudonym, if nothing else, indicates that.) But even hard-charging Clint recognizes moral limits to how the structure of agriculture in Powell County encourages him to act. As tight as things are, you don't economically stab other farmers in the back—at least you usually don't admit it.

But what's the trouble here, one might ask? Isn't the treadmill of production exactly what we expect and want capitalism to do? Through competition, economic efficiency is continually improved, leading to increased production and lower prices for consumers, as well as increased profits for those who are good enough to "stay in the game." Everybody benefits, or so it is often said.

It is true that the treadmill of production is often quite efficient in terms of money. But money is a notoriously poor measure of the things we really care about—community, the environment, and the long-term viability and equality of the economic fortunes of most of the people involved: in a word, sustainability. Farmers and those caught up in the treadmills of other industries become so overwhelmed with the individualistic struggle to stay on and stay in that these deeper concerns all too easily slip from view. And consumers themselves often get caught on their own treadmill, a treadmill of consumption, in which they seek to maximize their consumptive power and display.[27] What Thorstein Veblen long ago described as "conspicuous consumption" is a forever receding place to try to stand, as consumers try to keep up with each other, constantly accelerating this treadmill as well. Consumers readily become consumed by consumption, buying as much as they can for as little as they can, and they too lose sight of the deeper concerns of life.

Moreover, it is not clear that the treadmill of production always rewards efficient production. For the moment let's leave aside the question of whether effects on community, environment, and equity ought to be considered in the calculation of what is truly efficient production. Even viewing agricultural production in narrow terms of yield per dollar of input, the farms we lose are in fact often quite efficient. The long-established trend toward increased size of farms does not necessarily reflect some iron economic law that big is more efficient than little. On closer inspection, the situation is more complex. About all that one can safely conclude from the hundreds of studies on farm scale and economic efficiency is that there is no necessary relationship. Dairy farms are a popular basis for such studies, because the structure and technology of such farms are relatively similar across the industrialized world, and because dairy farms are less regionally specific than, say, wheat farms. Some studies find, as one might expect, that big dairy farms are more efficient producers than smaller ones. But a 1993 study of Wisconsin dairy farms found "important economies of scale on very small farms" and "some diseconomies of scale for the larger farms." A 1999 study of dairy farms in New Zealand similarly found midsized operations to be the most efficient.[28] It's just not a clear thing.

Gene Logsdon, the agricultural writer and small farmer from Ohio, remarks that "none of the small farms have ever been forced out by economics."[29] Not quite true. There is an economic treadmill. But what I believe Logsdon was trying to say is that it doesn't operate as advertised. Indeed, even the academic phrase "treadmill of production" misstates the primary reason smaller farmers are so often forced off it. Central to the farmer's problem is that the treadmill is actually based on the economics of money, not the economics of production. And this gives a big advantage to larger producers—not because they are more efficient producers, necessarily, but because they have the deeper pockets that enable them to ride through the hard times. My metaphor is rather mixed, but I hope the point is clear: The treadmill encourages monopoly power.[30]

Much of the monopolization happens during price swings for agricultural commodities. Farm circles are full of talk about "commodity cycles," as well they might be, given the structure of the farmer's problem. For example, between 1995 and 2000 the price farmers received for hogs went as high as 76 cents a pound and as low as 9 cents a pound. Breakeven is usually regarded as about 40 cents a pound for most farmers.[31] That means farmers with hogs to sell were making really good money at the top of the price cycle. But it also means they were absolutely desperate when hogs fell below a dime a pound, the lowest price since the Great Depression (and that's not taking inflation into account). Hog prices have shown an unusually wide range in recent years, but there have also been substantial swings in the price of corn, soybeans, and cattle, the other three main commodities for Iowa farms. The winter of 1998–99 was the worst, as all four hit the bottom of their cycles simultaneously. Thousands of farmers "went out."[32]

Moreover, those farmers who remain "in" find themselves under pressure to ally with monopoly powers through contract farming. It's an increasingly widespread practice in agriculture, and it involves farmers across the commodity spectrum, from conventional livestock confinement operations to organic growers. Although widely varied in their details, the basic formula of contract farming is the same: The farmer produces according to some specifications, and the contractor guarantees to buy at a set price what the farmer produces. This may sound like a good deal for the farmer, and it probably is when prices are low. But it leaves the farmer beholden to whomever holds the contract, with its many pages of caveats. Among other constraints, the contract usually requires that farmers produce using specialized machinery, seed, and buildings that the farmer pays for, often with loans from the contractor. The contractor thus passes much of the risk of such capital-intensive production on to the farmer, as well as gaining a second profit from serving as the banker for the farmer's loans.

The spread of contract farming also tends to shut out of the market farmers who don't have contracts. The processors and distributors prefer to deal with a few large entities, like the contractors, rather than many smaller entities, such as individual farms. Because they develop their business through vertical integra-

Silhouette of the elevators, water tower, and churches of Pocahontas, Iowa, 1989.

tion down the production chain (and sometimes up too), the contractors are often called "integrators"—a polite word for monopolists, say many.[33] It's an arrangement that bears an uncomfortable resemblance to feudalism, say some others, including former U.S. Secretary of Agriculture Dan Glickman.[34] Most farmers who invest $1 million or so in a hog confinement facility in order to sell hogs on contract know that they are losing their autonomy, becoming high-tech serfs of a sort. They also think they don't have a lot of choice, and they may be right.

Still, who cares about the farmer's problem? some might say. It sure keeps food prices low. But in fact this isn't so clear. Cheap farm production doesn't necessarily result in low prices in the supermarket. Oligarchic and monopolistic "middlemen"—these middlemen are corporations, not just a few guys with trucks—capture much of the price differential, resulting in big windfalls during dips in the prices farmers receive.[35] For example, when farmers were getting less than a dime a pound for their hogs, which is about 20 percent of what they usually get, there was no corresponding big sale on pork in U.S. supermarkets. (This little fact resulted in some grumbling in Congress at the time and some tentative efforts at action, including an investigation by the Federal Trade Commission, but with no real consequence.)[36] In other words, much of the retail pricing of food is done outside and regardless of market forces. As the old folk song "The Farmer Is the One" says, with only some overstatement, "the merchant is the one who gets it all." Finally, given that farmers get so little for what they sell—about 19 percent of the food dollar goes back to farmers—changes in their production costs have relatively little impact on the retail cost of food.[37]

The tendency toward monopolization is a widely recognized problem with capitalism, and it's something we look to government to hold in check, "to keep the playing field level," as people often say. But federal agricultural policy is notorious for doing just the reverse, in more ways than the much-criticized dairy herd buyout program. Of course, this notoriety is to some extent a matter of political perspective. But at the very least, it must be recognized that U.S. government policy has failed to halt the declining number and increasing size of farms during the perpetual farm crisis of the past fifty years and more. In Europe and Japan, by contrast, farm sizes for comparable kinds of farm operations are typically smaller, and often much smaller, which is largely the result of deliberate government farm and rural policies.

But monopolization is on the rise there, too, and many would point to the universal importance of the factors I earlier called "subterranean"—problems with family, health, management, and economic opportunity. These other factors, however, are not up to the explanatory task. True, individual farms face real problems with family conflict, health, poor management decisions, and simple bad luck in, say, guessing on grain futures. But there is no credible evidence to suggest that these factors affect smaller operations at a proportionally greater rate than they do larger ones. Big farms have family problems, too. The same goes for health problems. Some observers have argued that farmers who are able to increase the size of their operations must be better managers, and are better informed about new technologies and about how the farm commodity markets work.[38] (Indeed, this is a widespread belief among farmers themselves.) But such arguments are usually tautological: The common logic runs, if a farm is bigger it must be so because the farmer is a better manager and market watcher.

Far more significant in promoting monopolization in agriculture are the structure of federal agricultural subsidies, marketing options, and relations between farmers and bankers. Let's take up the last item first. What these relations show is the overwhelming importance of financial efficiency over productive efficiency. Farming today is incredibly capital-intensive for the return one gets. The land, buildings, equipment, and livestock alone can represent several million dollars in investment on a commercial-scale grain farm. Then there are the annual costs for seed, feed, fertilizer, pesticides, antibiotics, repairs, land rents, and more, which together can easily run into the hundreds of thousands of dollars. And yet returns in a good year are on the order of only a few percent. Thus the old joke: Want to make a small fortune in farming? Start with a large one.

This capital intensiveness does not necessarily reward the efficient producer. Rather, it means that farmers are forever beholden to bankers. And it's a lot more work for a banker to handle a big bunch of little accounts than a little bunch of big ones. Thus banks typically encourage farmers to follow the infamous advice of Earl Butz, U.S. Secretary of Agriculture in the mid-1970s: "Get big or get out."[39] After the farm crisis of the 1980s (which many have argued was brought on in part by advice like Butz's), bankers got a bit cautious for a while. When the market for farmland went sour, the banks called in a lot of loans, forcing thousands of Iowa farmers out of business. But the prices for farmland have picked up again in Iowa and elsewhere in the grain belt, and the banks are once again encouraging farmers to buy and rent more land, forcing still more thousands of farmers out of business. This is why the 1980s farm crisis barely registers on graphs of farm numbers. In bad times or good, the land market forces farmers out.[40] When the agricultural economy is weak, many farmers are forced to sell out. When the agricultural economy is strong, farmers and their bankers have plenty of money to spend going after more land. As Clint explains, "It used to be, the bank was real tight with cash flows. Well, they're getting to be a little bit more, 'you want to buy it, go ahead'—even if it's a bad deal."

Because either way it's a good deal for the banks: fewer, but bigger, accounts to manage.[41] Consequently, the entrance bar to farming gets higher and higher, and fewer and fewer young people see a future in it. Each year the average age of farmers goes up. It's currently 52.4 years in Iowa.[42] Not only are we losing farmers, but we are discouraging young people from even becoming farmers—which is both an economic and a cultural threat to rural Iowa and places like it. In fact, there has been little change in recent years in the rate of what rural demographers call "farm exits"—farmers leaving the land because of retirement or financial troubles, or for other reasons. What has really changed is that there has been a dramatic decrease in the rate of "farm entrants"—people starting out in farming.[43] So much capital is needed to put yourself in the unenviable situation of working hard under conditions of significant uncertainty in order to make at best a modest income that even children raised in farming traditions are looking elsewhere for a livelihood. Their parents increasingly concur. One bitter line we heard several times during our field work is that "encouraging your kids to go into farming is a form of child abuse."

Clint wasn't so bitter about it, but he was certainly negative about the prospects for the young aspiring farmer. I had asked him, "What would be your most important advice to give somebody else starting out in farming?"

"Probably not to do it," he said with a rueful laugh. "Because I don't think they could unless there was somebody, their father or somebody else, that would want to take them under their wing. Otherwise I don't think they could do it. Maybe I'm wrong to think that. But what it would take investment-wise, and the money it takes, I don't think you could just go to the bank and get money without somebody co-signing it. Just couldn't do it."

The banks face their own constraints as well. If the farmers who remain are increasingly beholden to the banks, the banks themselves are increasingly beholden to the farmers who remain. As Iowa State University agricultural economist Mike Duffy puts it, "if you owe the bank a hundred thousand dollars, you've got a problem. But if you owe the bank a million dollars, the bank has a problem."[44] What Duffy means is that it is easy enough for a bank to foreclose on a farm that's in debt for $100,000. The bank may lose a little, but it will also then have the opportunity of investing that money somewhere else—probably in a bigger loan to a bigger farm with lower administrative needs per dollar invested. The bank may even come out financially ahead on the deal. But a default on a farm in debt for a million dollars is a lot bigger thistle for a bank to swallow, as it grazes across the rural economy.

What this means is that if a big farm runs into difficulty—whether for reasons of family, health, management, or market troubles—a bank is far more likely to step in with yet another loan, letting the biggest gambles it has made ride far longer. Even with a big farm, of course, at a certain point a bank is likely to stop throwing good money after bad. But, at least in farming, there is much truth to the aphorism that "the more in debt you are, the easier it is to borrow money," and we heard several farmers use it.

And then there are the federal agricultural subsidies—well intentioned, perhaps, but paving stones to agricultural purgatory for small farms nonetheless. Iowa farmers have been receiving more than $2 billion in subsidies in recent years—1999 to 2001—or about $22,400 per farm.[45] Admittedly, the average from 1996 to 1998 was about $0.8 billion, or some $8,300 per farm.[46] Still, that's a lot, an awful lot, and it should have kept just about every Iowa farmer in business. But this kind of federal fertilizer is spread overwhelmingly on a few farms, typically large farms specializing in grain production—what are termed in agricultural circles "cash-grain" farms (with unintentional aptness). According to an analysis conducted by the Environmental Working Group of the federal subsidies Iowa farmers received under the Freedom to Farm Act from 1996 to 1998, 44 percent of the subsidies went to just 10 percent of the state's "farm entities" (that is, both family farms and large agricultural corporations, as well as other forms of farm organization). The leading recipient averaged more than $200,000 a year; the top seventeen recipients took in more than $100,000 a year. (And this was before subsidies nearly tripled after 1998.) Meanwhile, the bottom 21 percent got $1,000 a year or less, and 10 percent averaged just $200 a year or less.[47]

Paradoxically, subsidies are usually defended in Congress as support for the family farm. The problem is that rather than doling out subsidy money on the basis of income adjustment or environmental management—that is, on the basis of real needs or good deeds—most payments are based on acres in a few standard crops, such as corn, soybeans, wheat, rice, and a handful of others. Strictly speaking, after the passage of the Freedom to Farm Act in 1996, most subsidies were based on a farm's history of acreage in the subsidized crops. For six years there wasn't a requirement to maintain one's "base acres" in the subsidized crops; you could plant something else and still get the subsidy, once you had registered what your base was as of 1996. That's what the "freedom" in "freedom to farm" meant. The 2002 farm bill, the Farm Security and Rural Investment Act of 2002, allows farmers to continue with the base they established in 1996 or to calculate a new base using the actual acreage in the subsidized crops that they planted the four years previous to the 2002 bill. Confused? So are many farmers themselves. Farmers who were good with the crystal ball and actually planted more acres in the subsidized crops during the freedom-to-farm years (rather than less, as many farmers thought was being encouraged) have thus been able under the 2002 bill to increase their base and therefore their claim on subsidy dollars.

What all these arcane rules mean is that farms with a strong history in the subsidized crops are attractive investment options. These heavily subsidized farms are most readily available to those who are able to pay for them, or to get the loans to pay for them: those who have big farms already, and that means primarily specialized grain farms. But most of these farmers are already getting a hefty slice of subsidies (which is one of the main ways the big farms got so big). Consequently, "the current programs nail the little guy," as economist Mike Duffy pithily put it in a congressional hearing.[48]

Moreover, the big farms are usually able to get more for what they raise. One of the main caveats that Adam Smith set out in defense of capitalism in *The Wealth of Nations* was the principle of open access to markets. But purchasers of grain and livestock are in much the same position as banks. They too would rather deal with fewer, smaller accounts to reduce what economists term "transaction costs." The vertical integration in agriculture that I described earlier gives big farms better access to markets, and better prices with that access. One study found that large farms received 12 percent more for their corn and 16 percent more for their soybeans than smaller ones in 1997, and better market access seems to have been why.[49]

What Earl Butz should more accurately have said is, "get in debt in a big way or get out." That way you should be able to gain control of lots of acres and thereby more readily command an ample portion of federal largess, at the same time putting yourself in an advantageous marketing position and gaining the commitment to your survival of banks, markets, and other agricultural institutions. For given the pervasiveness of the farmer's problem, the agricultural race is not necessarily to the swift and sleek but, in most cases, to the large and overextended.

2. Community and Environment

More than individual farmers get squeezed out by the farmer's problem. Whole communities do. The changing structure of agriculture is restructuring rural culture, as the previous chapter describes. It is also presenting new challenges to rural people's communal ties with each other. It is hard to maintain a community based on agricultural businesses if there aren't many farmers. And it is hard to maintain a community's fabric of trust and fellowship if people find themselves rending it more than mending it.

Clint understands the economic dimension of this clearly, as do most farmers we visited with during the field work. He and I had been talking in the kitchen about his ambivalence about whether his own children should go into farming, and his feeling that at the very least, unlike him, they should get a college degree first, so they have more options than he had.

"So, do you feel optimistic or pessimistic about rural life and farming?"

"You're gonna have to be aggressive to stay in," he replied, starting with the first principle of life on the treadmill, as he sees it. He took a last sip of his coffee. "But I'm a little pessimistic about in town. It don't take a rocket scientist to figure out there's gonna be less people here. As for the ag businesses [in town], a lot of them are gonna go to the wayside too, I feel. Because chemicals, I think they're gonna be in a lot of warehouses. A lot of chemicals are gonna come UPS."

Even the agricultural chemical dealers in town may soon be out of business, Clint was suggesting, as the agricultural supplies market shifts to remote warehouse merchandisers. "Order that stuff right through the computer," I added, making sure I was understanding him.

"Right. [Consequently], like in Powell Center, there just won't be the jobs. The kids, if they do get a college education, most likely aren't gonna be able to come back and live in Powell Center. Unless we get some sort of a something going here, there isn't anything you're gonna be able to do. You know, to utilize your education. So, I'm a little pessimistic for the town. It's sad, but," he lifted his cup for another sip. Then Clint set it down before bringing it to his lips, remembering he'd already drained it.

"And there's just gonna be less and less farming," he went on. "That's just the way it is. The equipment and stuff." He raised his right hand and rubbed his fin-

Town welcome sign, Vinton, Iowa, 2000.

gers together in the international symbol of filthy lucre. "I know of a lot of people who are looking to farm more ground. And like I told you, I'm not gonna run out and try to cut throat, but I would—"

"If a farm comes up."

"Yeah, I'd try to get it. And you just drive down the road here, and I bet most of them would tell you that. They would rent a little more, unless they own their own and don't rent any. It's scary, the changes there's been in livestock in the last five years. Look what can maybe happen to the hog industry."

Has happened. Nowhere has farm concentration been more pronounced in recent years in Iowa then in hog production, with the coming of the big hog lots. And that spells harder times for Iowa's small-town Main Streets as well.

"There's gonna be a lot of changes. And if you're not aggressive or—" He shook his head. ". . . a lot of people are gonna retire, and a lot of these people are wanting to rent their ground. So, it is sad, [but] that's the way it is."

The theme of aggressiveness in the face of economic competitiveness is, of course, a familiar one in a capitalist society. It's often celebrated in contemporary culture, and Clint too takes pride and comfort in it, finding self-respect in his aggressiveness and reassurance in his sense of its inevitability. Clint doesn't see a contradiction between aggressiveness as something to respect and as something that is inevitable, and neither does American popular culture. Rather, Clint seems to see in these ideas justification for his economic aggressiveness. As it's inevitable—it's sad but that's the way it is—he can't be blamed for it. As it's an achievement worthy of respect—you have to be aggressive to stay in, and Clint has achieved that thus far—he can nevertheless take credit for it too.

Still, Clint recognizes an issue that the common celebration of economic aggressiveness usually overlooks: that it's cold comfort for the erosion of community life. Crocodile tears, perhaps, but there may be something about being a farmer in a small Iowa town that makes the potential benefits and pleasures of a vibrant local community more readily apparent than is typical in America today. It really is a long way to the neighbors', to shopping, to the nearest hospital, church, and school, to off-farm work—and increasingly to farm work too, as farmers rent more and more ground farther and farther from home. Even with modern transportation, electronic media, cell phones, the Internet, and the rest of it, distance does still matter in human relations, despite many decades of hype about how technology now makes possible globalization, virtual realities, and "community without propinquity."[1] Moreover, people in rural communities don't have as wide a range of folks to draw upon for their business needs and contacts. If a local business goes under, it may easily mean going twenty or thirty miles to find a similar service.

But whether or not I am right that rural living tends to make plainer the value of local community, we were struck by how farmer after farmer we spoke to recognized the implications of their farms for the local community and economy, and vice versa.[2] Not all to the same degree. Not all expressed the same level of concern about it. But, like Clint, they all readily understood what we meant when we brought it up in interviews, and they often brought it up themselves.

Call it the "plywood effect." It's hard not to think about what's going on in the local community and economy when the stores in your town start getting boarded up. You may or may not be much troubled by it. You may not understand it fully. (Indeed, no one does, given the state of our understanding of community and economy.) But when the evidence is so close at hand, you are nevertheless likely to reflect on why there's no longer much glass in the store windows to reflect from.

The effect of agricultural structure on community is not just a matter of economic vitality, though. It's also a matter of the quality of human ties, as a number of farmers pointed out to us. In rural Iowa, neighbors mean more than what or who is adjacent to some place. They mean a social relationship. People talk about going out "to neighbor" with the family next door. They talk about the importance of "neighboring" with others—that is, helping out in times of need, and also for pure sociability. Often that next door is some distance away, which may be the origin of the idiom "visiting" to describe the act of having an intimate conversation. But a common complaint that we heard about current times is the decline in neighboring and visiting. Wendell said it this way, stumbling to find the words for what troubled him.

"I think that is probably one of the things that has changed drastically in agriculture. I know there's different ones that talk about it and still think of neighbors. But I don't know. It seems to me that there's some of them that are just—I don't know, maybe it's just me. There is a little bit of competitiveness. It's not like it used to [be]."

He pushed up the brim of his hat, seeming to consider whether he himself had ever experienced things as they "used to be." Then he continued, "Well, you got to go back quite a ways. But you used to do more neighboring, and used to help one another out. I'd like to see that come back again. Maybe that's going to be one of the ways of a smaller farmer being able to survive with the high-priced machinery and stuff. We're going to have to get to where we neighbor more, you know, like we used to do. . . . [But] I do kind of feel that, maybe with the higher price rents and the way things are, these larger farmers, they're not really neighbors."

Dale farms with his son, near Wendell and Terri, and I interviewed him the next day. Here's how he put it.

"We do neighbor some with our guy south of here. We work with him some. We use some of his machinery and then we help him out. He feeds a lot of cattle, so we go down there and kind of work with him some. But, yeah, 'cause most of our neighbors are grain farmers and with different philosophies and ideas—I don't know. You just don't feel comfortable around them like you should. I mean, it isn't like back thirty years ago when everybody basically had the same thing and did the same thing and were basically in a survival mentality. Then everyone had a tendency to work together more. Things have gotten to the point where the more self-centered—"

Dale stopped for a moment, perhaps deciding that "self-centered" was too negative a term for an interview situation. Then he went on, "where everything is done basically for yourself or your self-betterment. So you've kind of lost that need for that neighboring or getting together and talking about things in common."

What Wendell and Dale are talking about is something that many farmers do in fact still do: share equipment and trade chores, just like neighbors in suburbia will borrow a saw or help out with each other's home improvement jobs. There is, of course, an economic dimension to this borrowing and exchange. As Wendell suggests, it might be a way for smaller farmers to survive economically. You buy the rotary hoe, I'll buy the Buffalo planter, and we'll share each other's equipment and save money. But this kind of nonmonetary economic exchange also contributes to a feeling of being "comfortable" with one's social surroundings, as Dale says—a sense of safety, perhaps, that comes from working together and knowing that not all relations have been reduced to the unforgiving metric of dollars and cents.

Of course, there is much scope for romanticism here, as with the notion of the "loss" of rural culture that I discussed in the previous chapter. There was a lot of competition in the past too. People often overstepped the community's sense of the bounds of decency in their efforts to get ahead economically. Indeed, people everywhere have long complained of a decline in human cooperation, as far back as we have records of human complaint—the "golden escalator" again.[3] Complaining about a decline in community is in fact one of the ways that we mobilize community feeling and cooperation, needling each other with a touch of guilt.[4]

But it is also true that it is harder to mobilize that feeling and cooperation when the interests and orientations of those in our community are increasingly different. Dale noted this, but Elwin, another Powell County farmer, put it even better.

Elwin is forty-ish, a friendly man with a broad grin. On the day I first visited his farm, he was wearing a green and white nylon cap—worn and dirty, and clearly much loved—that read "New Farm" across the brow. "New Farm" is the name of the Rodale Institute's Internet magazine devoted to sustainability and small-farm agriculture. Elwin is an avid reader. His full-time farm is only about 450 acres—which is small in the view of most full-time Iowa farmers. He also has a hog operation, raising some thousand head a year in "hoop houses," an increasingly popular husbandry method on small farms.

A little aside on hoop houses: These are basically just big nylon tarps stretched over a framework of metal hoops, resulting in a structure some seventy feet long, thirty feet wide, and twenty feet high, aligned north-south, usually open at the front and back or closed at the north end in bad weather with another nylon tarp. Farmers fill them with deep beds of straw and let the hogs nest and root, much as they would in the wild. The straw mixes with the animals' manure, absorbing the fluids and aerating it at the same time, thus keeping hog odor down. The straw-manure mix also composts under the hogs' feet, providing a significant heat source in winter. Some suspect that the compost burns up pathogens year round, helping keep the hogs healthy without antibiotics.[5] Neighbors like them because of their reduced odor. Hogs seem to like them too; they fight less and play more than in hog confinement buildings.[6] And small farmers like them. (Some bigger farmers, too.) Not only are hoop houses pleasant, airy places to work. Not only are their animals and neighbors happier. Not only are they easier on the environment. They're vastly cheaper to build than conventional facilities—about a third per animal in construction costs. They do require a bit more labor per animal to manage.[7] But Elwin has more control over his labor supply than his money supply, and hoop houses have thus become popular among small farmers like him, particularly those attuned to ideas of sustainability.

"Mike? How ya doing?" Elwin called across the farmyard to me, as I stepped out of the Iowa State University car that I had ridden down in. Elwin waved me into his cluttered farm office in the back of a big gray machine shed at one side of the farmyard. Although he farms full-time, Elwin also maintains a sideline selling Hill Seeds, a local Iowa brand often advertised during Iowa State football games. Much of the office was given over to a metal conference table with a display on Hill Seeds set up for his sales work. There was also an extensive array of computer equipment, the monitors piled high with magazines and covered with sticky notes. Next to one monitor was a stack of paperback books. One of the titles caught my eye: *A Sand County Almanac.*

We began with Elwin looking over a book of my own, a work I published a number of years ago on social class in an English exurban village. Although he was cordial, I think Elwin was, unsurprisingly, also a bit suspicious of an East Coast sociologist come to ask about his views on farming. As I sometimes did, I pulled my earlier book out of my backpack to give him some idea of what I was up to. He read the blurb on the back cover and flipped through the inside while I cradled the Styrofoam cup of coffee he had poured for me. (It was cold in the shed.)

Then he put the book down and shook his head. I had a flash of concern about Elwin's reaction to my book, but that's not what he meant.

"What I'm worried about is that in the future Iowa is going to see more class division," he said.

I nodded and replied, "As you may have read in the *Des Moines Register* this fall, the average wage in Iowa has fallen at the same time that the amount of wealth in the state has grown."[8]

He nodded in return, and picked up a pencil and began to draw. "Iowa used to be like a pot bellied stove," he said, drawing a surprisingly decent picture of one. "And now—you can probably guess what I'm going to draw—" He scribbled a bit more. "And now it's going more like an hour-glass."

He kept drawing. "Or maybe more like this. I don't know what you'd call it."

"A flask," I said, looking at the shape he'd drawn.

Elwin nodded. "The way I figure the hog business—really, any business—is that you usually work for minimum wage, or less. Then comes the windfall. That's what gets you through." An excellent description of the farmer's problem, I thought, and how to get through it. Then Elwin went on to add another dimension. "What happens when Murphy comes in is the windfall goes to them, and out of the community."

Despite its rustic name, Murphy Family Farms was one of the big corporations that brought large-scale hog confinements to Iowa in the 1990s, sending hog farmers out of business and paying workers in the new facilities wages little above those paid in convenience stores. (In 2000, Murphy was acquired by Smithfield Foods, the world's largest pork processor, in a vertical integration move that eventually made Smithfield also the world's larger pork producer.) The convenience stores are also a problem, as Elwin went on to note as well.

"It's the same with Casey's," he said, tossing his pencil down on the table and pushing back in the old gray metal swivel chair he was sitting in. "The people work for minimum wage, but the windfall goes to Des Moines."

Casey's convenience stores—headquartered in the Des Moines suburbs—can be found all over rural Iowa, scooping up a lion's share of the local trade. Plenty of small Iowa towns no longer have a functioning Main Street, but few are ungraced by a Casey's or a Kum 'n' Go or a Pronto. Very often it's the only retail business left.

"And that leads to this," he said, pointing to the flask shape. Elwin picked up the pencil again and returned to his earlier drawing, adding in some circles in the middle of the pot-bellied stove as he spoke. "Here's where the church committees are, the farmers, the merchants are. These are the people who get things done in a community, and they're getting squeezed out."

"So you mean that as the people in the middle get squeezed, we lose community?"

"Exactly right."

"You're a pretty good sociologist, Elwin," I said. And I meant it.

Pronto gas station, convenience store, and coffee shop, the retail hub of Rolfe, Iowa, 1996.

But it's more than the people in the middle of a rural community that get squeezed by the structure of agriculture and the farmer's problem. It's the people on the bottom too—as well as animals and the land.

The reach of the squeeze down the human community and out of it into the broader ecological community was very apparent one morning to Greg and me when we visited a midsized hog confinement operation in Powell County. After a couple of wrong turns, we finally found the place: IQP—Iowa Quality Pigs, Inc., as it said on a modest yellow and black sign by the gate.[9] Greg and I had been eager to see the inside of a hog confinement operation, and a farmer Greg had interviewed had offered to arrange a tour for us with a friend of his who worked at IQP. Greg had called the friend, Brent, a few days earlier to set up a time and get directions. But something got lost in translation, and thus so did we, for half an hour or so. Given the controversy surrounding them, it's not easy to arrange a visit to a "slat-barn" hog operation, so we were quite concerned about being late. We didn't want to blow the opportunity.

We roared up the gravel drive. The site was dominated by half a dozen long metal barns, a couple of them connected to a central building, which had a small sign outside saying "office." Greg pulled his pickup into a space by the office, and we jumped out. A few other vehicles were parked in the muddy lot, all with license plates from nearby counties, but not Powell. (Iowa, like many Midwestern states, prints a vehicle's county of registration on the license plate.) We couldn't see or hear anyone, but the office door was open, so we walked in.

Actually, only the top half of the door was open. It was one of those divided affairs one sometimes sees in the offices of industrial facilities, where the top half of the door can be opened independently of the bottom. We didn't see anyone within, but there was a small waiting area just inside with a dusty-looking broken-down couch. That was invitation enough, we thought. There was no handle on the outside, so Greg reached over and opened the bottom half of the door, and we went in.

We didn't sit on the couch, though. There was just too much dust—dust everywhere, on the little dorm-room-sized rusty refrigerator, on the Mr. Coffee coffee maker balanced precariously on the portable television, on the beat-up filing cabinet, on the desk with the telephone wrapped in plastic, evidently to protect it from the dust. Hog confinement operations are notorious for their dust, a mixture of hog dander and fine feed particles, and we could readily see why. "Rotate Semen Twice A Day" said the sign on the fridge. In the back of the office was a door with another sign: "ABSOLUTELY NO ADMITTANCE! All Visitors to IQP Must Sign In Saying They Have Not Visited a Hog Facility in the Last 24 Hours." Disease is a constant worry when animals are raised at such close quarters, and this door evidently led to the shower facility, part of the operation's efforts to control the spread of pathogens. Masking-taped to the wall by the door was the sign-in sheet, a torn piece of yellow paper with roughly twenty names and dates going back about two years. Another door, which was ajar, led to a bathroom; over the toilet we could see a gun-rack with three big shotguns. There was also a toaster oven and an open box of face masks on the shelf next to the fridge, and a cracked Styrofoam cooler, probably for transporting the semen. A half-finished quart-sized plastic bottle of pop stood open on the desk.

We stood there in a daze, slowly taking in our surroundings. Finally Greg said, "Mike, I know there's that open bottle of pop. But I think we could be waiting a long time. Besides, we're late. What do you say we look around and see if we can find somebody?"

I nodded, and we stepped back outside. The smell of hog manure was everywhere. We ducked down a narrow alley between a couple of the metal barns, thinking we might find someone in back. For some reason, halfway down the alley a bucket was set up at the base of a downspout. A used plastic syringe floated in a couple inches of water at the bottom of the bucket. At the end of the alley we came on a pile of bones, pig, we assumed, mainly jaws. An old oil filter for a truck or a tractor sat off to the side in some mud. Just beyond lay two vast manure lagoons, open to the sky. I thought about the story I'd read recently in the *Des Moines Register* of a large manure spill from an operation like this in another Iowa county. Thousands of fish died, and the local ground water almost certainly received a big charge of nitrates. In the six years from 1996 to 2002, there were 152 spills from Iowa hog and cattle manure lagoons, killing an estimated 5.7 million fish.[10] Just beyond the lagoons, the waves of corn and soybeans returned. A small farmhouse lay off in the distance.

We heard some echoey voices but didn't see anybody. Greg raised his eyebrows at me. "Spanish," I said, finally making out a few of the words. Greg pointed to a metal grate high in the wall just above our heads. Evidently we were hearing workers talking in the adjacent building, their voices carrying through some kind of vent. I tried to think up a Spanish salutation appropriate for the situation, but my Spanish isn't great and before I could formulate a sentence the voices got quieter and drifted away.

"Look, Mike," Greg said. "Why don't you go back and wait in the truck in case Brent shows up. I'll keep looking around and see if I can find somebody." I headed back to the truck and amused myself by counting the number of country radio stations you can pick up in Powell County. (Five, it turned out, at least that afternoon.)

Ten minutes later the truck door on the driver's side opened and Greg threw himself in, a bit breathless. "Brent's not coming. He had to go to a funeral. A worker told me about it. And I got a good look inside. What a place! Let's go." He started the engine. "Turn the tape recorder on and I'll explain."

"Yeah, it was something," he began, when I switched the tape recorder on. "I'm walking around, just wondering, you know. I'll case the joint and see if maybe wherever the manure comes out maybe these guys walk out. . . . You got that going?"

"Yeah, I got it going."

"So I was walking around the back of the building. There are feed hoppers back there, and I saw the auger things going down in them. And I could hear the hogs. Maybe we can just sit in here," Greg said, pulling the pickup into a little rest area just out of sight of IQP. "I don't want to talk and drive at the same time."

"All right."

"Then I saw a garbage heap back there and I thought, hey man, that's where that smell is," Greg continued. There had been a terrifically acrid smell mixed with the hog odor as we had walked around the IQP barns. "There was another little alleyway, and it was all full of crap. It was all full of mud. There were some dead pigs that were all black and nasty, decomposed, laying on this little pile. And there were old cages, old racks."

"This is in the back side."

"Yeah, back side."

"They don't throw their dead pigs on the front side of the building. This is clear."

"That's right, it was the back side. The pile was up against the wall like they'd just thrown them to get them out of the way so you don't have to walk over them or whatever. There were tractor tracks back there so maybe they haul them with that tractor, maybe haul the carcasses out with that. . . ." There had been a big John Deere tractor parked behind one of the buildings. "Anyway, I saw this door, and I walked up to it."

The door was shaking and Greg saw a steady flow of gray water running out at the bottom of it with little granules of feed and bits of concrete. He could hear the sound of a compressor and water spraying inside. Greg pounded on the door, and after a moment the compressor went quiet and the door opened. A man who had evidently been power-washing the inside of the barn looked out. Behind him Greg could see metal cages full of hogs, a maze of hoses for the automatic feeding and watering systems, the cement floor the man had been washing down, a big sheet of insulation falling out from the ceiling, a ripped-up plasterboard side wall that was wet and falling apart, and a few incandescent bulbs glowing dully in the gloom.

"Big guy, maybe five-ten, but pretty stout," Greg described. "He was wearing a button-down shirt. It was blue, kind of nasty. He had hip wading boots that were black and rubber and he was just full of crap. He had a beard and his beard was all full of crap. Crap all over him. And in his hair and covered all over his clothes. You know, the power-washing deal. Well, he comes barreling out of there and says, 'Nobody home.'"

It was an inauspicious beginning to an impromptu interview. But the man took the opportunity for a cigarette break and proceeded to visit with Greg about work life on a modern, high-tech hog "farm" (as Brent had described the facility to Greg on the phone). It helped that Greg had at one time worked in a similar facility, and they quickly established some rapport.

"This is the job," the man said to Greg with a bit of bravura, pointing back at the power washer. "It isn't so bad. But you get real dirty, real messy."

He took a deep drag on his cigarette. "But it's not the job that you'd want," he went on. "Those fucking Mexicans will fight over getting this job, so I told them to get somebody like that up here so we don't have to do this job. We all hate it in here."

Remembering the Spanish we had heard drifting through the vent, Greg wondered exactly who was this "we" that seemed to exclude the Mexican workers at IQP. (The "them" was clearly IQP's owners or managers.) Greg tried a neutral question. "So you guys aren't power-washing every day?"

"Exactly, because we hate it. What we do is rotate between the four of us. So [I said] 'you get one of those Mexicans to come up and take care of it.' Because we all know that they like to do that dirty work."

Greg gave his best yuk-yuk laugh and hoped it would pass for real. There may have been resentment behind this racist remark. Perhaps the man felt aggrieved that even though IQP evidently had Mexican workers at this site, he was nevertheless still doing the power-washing. Maybe in the presence of another white man he felt embarrassed about it. But there was no polite way to ask about any of this. So, trying to find some remark that would connect honestly, Greg mentioned that at the place he once worked there was always a bit of struggle to get someone else to do the power-washing.

"Yeah, here too. The new guy always has to do the power-washing job. . . . The other thing we make them do is help with the births," the man went on. "If they can't handle that, and that's too much for them, then they're not going to make it in this job. . . . That reminds me of a guy we just had. He was here for a while. He was almost here for a year. And then we had to do the Cesarean section on this hog. That guy ran out and he never came back. . . . You know, some of the smaller operations will make you do all of this stuff and you only make sixteen thousand a year."

This man probably wasn't making much more than that, but he seemed keen on giving Greg an idea of his salary without being too specific, perhaps again out of an effort to make a good impression with another white male. "Yeah, it's kind of hard to bust your butt for sixteen thousand a year," responded Greg.

"Exactly," agreed the man, and rubbed out his cigarette on the muddy ground.

Cut. It is springtime in rural Iowa and, as Rachel Carson predicted, it is silent, dead silent. My family and I are driving through northwest Iowa in mid-April, taking the back roads, as we love to do. We have been driving for nearly three hours, so we pull off for a minute to stretch and look at the map. No bird sings. No insect buzzes. Nothing moves on the landscape except a tractor in the distance, pulling a big spraying rig and trailed by a billow of what we hope is mostly dust. There hasn't been much rain, and the landscape is as drained of color as it is still, except for the grass greening up along the edges of the fields. The view seems much like Dorothy's Kansas in the first scenes of *The Wizard of Oz*, except the filming is in brown-and-white instead of black-and-white. Virtually everything—soil, crop debris in the fields, buildings, the gravel roads that peel off north and south every mile—seems to be some shade of brown.

We notice on the map a small prairie preserve, and we decide to head that way to take an extended break from the driving. Iowa is virtually all private land, and 92 percent farmland, the highest percentage of any state, so we are pleasantly surprised to find a small park close at hand. We barrel along, caught in the untornado that these shotgun-straight roads are, which force your car to keep to the rectilinear grid of blacktop and gravel, despite any desire you may have for a curve, any curve at all. Our car is Dorothy's house, I muse, out of control in a rushing monotony of brown silence.

At the sign for the preserve, we turn up a gravel road into a more rolling terrain. Our windows are closed to keep out the dust, but we hear something above the rumble of the wheels on the gravel. We look at each other but can't quite make out what it is. A small bird flies a low roller coaster of a course right in front of the car. Suddenly we recognize the sound: bird song, muffled but, quite amazingly to us, loud enough to penetrate the noise of the car and the road. We open the windows to hear better and look off to the left, the direction in which the bird has flown. There is a section of land in tall grass, and it is alive with bits of colorful movement, the little looping flights of bird and insect in and out of the spiky tangle of the prairie meadow, for that is what this is. We pull the car in where a second sign indicates and step out into a roar of life: the real Oz.

Early one morning, I walked out alone
Looked down the street, no one was around
The sun was just coming up over my home
On Hickory Street, in a little farm town

Early, my Home on the Range
It's a one horse town, but that's alright with me

Cut again. We've left the prairie preserve more than an hour behind, but it's still another hour at least to Ames, where we're headed, and we're singing Greg Brown songs to keep us going. Greg Brown is a singer-songwriter from Iowa, for many years a regular on *Prairie Home Companion*, and several of his songs

Main Street, McCallsburg, Iowa, 2000.

are perennial favorites of ours. "Early," a lovely song he wrote about the town of Early, Iowa, is one of his best known. We notice that by chance the route we've chosen takes us right past Early, in just a few miles, and we decide to stop and have a look.

Early turns out to be a classic Iowa small town of wood frame houses, a few churches, a grain elevator, a water tower, and a population of maybe six hundred. We head for the main street. It's wide and grand, but everything looks boarded up. No one's around. We decide to hunt up Hickory Street, the one mentioned in Greg Brown's song, and start wandering Early's small grid of roads. Most of the cross streets have tree names: Maple Street, Oak Street, Walnut Street. But for some reason we can't find Hickory Street. We can't find anyone to ask either. No one's around.

We head back to the main street to see if we can find someone there. As we do, a pickup truck pulls up at a little building that we now recognize as the post office. Somehow we had missed it before. It's open, and the driver of the pickup goes in. I head in too and wait my turn. When it comes, I ask where Hickory Street is. "Who are you looking for?" comes the immediate reply from the woman behind the counter. I explain that we're not looking for anybody, but we just want to see the street mentioned in the Greg Brown song about Early. She has never heard of the song or Greg Brown. But she directs me to the area of town where we had just been.

I thank her and go back out to our car. There's no one around again, other than us. We give Hickory Street one more try, with greater confidence this time, and this time we find it. It's pleasant, lined with trees, with small, early twentieth-century houses, most with front porches. We look down the street to see if there's someone we can ask about the Greg Brown song, and about whether he ever lived there. But no one's around. Like the song said, no one's around. It's a silent spring here too.

3. Home and Family

The squeeze of the "farmer's problem" is hell on home life, too. Iowa farmers and their families feel themselves always rushing, rushing, rushing, caught between the demands of the structure of agriculture and the needs of family members.

The tension between work and home creates a very American dilemma, of course, one that is by no means limited to farm families, and a number of recent studies explore these tensions.[1] Increasingly, we are all caught in a "time bind," notes the sociologist Arlie Hochschild in a book by that name, and with the apt subtitle "when work becomes home and home becomes work."[2] But while I know of no study that compares hours of leisure time among farm families in contrast to other families, there is something qualitatively different about the time bind that farm families face. The nature of the farm enterprise is such that home and work are almost unavoidably deeply mingled. The sociologist Christena Nippert-Eng describes how contemporary American families make use of "segmentation" and "integration" strategies to cope with the home-work boundary. For farm families, segmentation is largely unattainable and integration is largely inescapable, making these strategies of limited use for them.[3] Thus the social consequences of the time bind strike farm families with special force. Moreover, because of the image of farm life as a refuge from the "rat race" (with its treadmill imagery), I couldn't help feeling at times that some farm families almost feel betrayed by the reality of their lives.

It's the small things that really matter here, the events that pass unattended and the tasks that pass unattended to. Farm men often spoke of their inability to get to a child's sports event, especially during planting and harvest seasons. "June is my stress month," Earl told me one rainy morning as we stood around his farmyard in late spring. Earl and his wife Kerry farm about 350 acres, largely organically, and raise free-range hogs without antibiotics. He's normally lively and energetic, but not that day. It had been raining a lot that spring, and he still didn't have his beans all planted, even though it was mid-June.

"This is a hell month, a month of hell." He shook his head, disconsolately. "It reminds me of the way I felt last November." His daughter Raylene had a softball game during the height of the last harvest season, he explained. "I was sitting

Connie, Johnny, and Mark Tjelmeland, PFI *farmers, at their farmhouse near McCallsburg, Iowa, 1998.*

there at that game and I just couldn't stop thinking about the farming. You know, that made me really depressed. I couldn't even enjoy her game. It took me two months to get over that one."

At least Earl made it to the game. Floyd finds that he rarely can manage even that. Sue and I got the impression that he doesn't try as hard as he might, though—at least, this was very much the view of Floyd's wife, Cheryl, as we learned one evening at their home. Sue and I were doing some interviewing in their area, and I had interviewed Floyd that afternoon and he invited us to stay for dinner. We each had another interview to do that evening, and the only nearby eateries we knew of did not look like the kind of places William Least Heat Moon would give many stars to. Besides, it was a nice offer. We accepted gratefully.

Cheryl did not seem shocked or annoyed when Floyd informed her, after the fact, that he'd invited us to stay for supper. Indeed, she seemed enthusiastic about it, and in short order whipped up a dinner for six (they have two children, a boy and a girl, both in their early teens) of beef steak, frozen corn, green beans, and two kinds of pickles, all raised on their farm or in Cheryl's garden. It was an impressive performance, and they both delighted in telling us about how they raised each item, and how they like to get together with various local relations for canning parties. The corn came from one such party last fall when, along with Floyd's brother's family, also farmers, they did up two hundred bags. They also spoke at great length about the importance of their extended family and how committed they are to it—as well they might be, given that their conventionally managed sixteen-hundred-acre farm is a partnership with Floyd's brother's family—even to the point of putting out a family newsletter a couple of times a year in which all the branches of their families share their latest doings. Despite the busy-ness of their lives, they always make time for these homegrown activities, they told us.

It did seem to us as a bit of performance, though—as perhaps inescapably it was, given who Sue and I were and why we were there. But it is worth pointing out that throughout this performance Floyd's only role was to talk. He didn't lift a finger to help Cheryl put out this dinner and he saw no reason to do so, despite the performance-like quality of the meal. Indeed, it seemed to us that his lack of involvement in the kitchen was part of the performance.

Or rather the performances, plural, for it emerged that there were several going on. Floyd seemed to be telling us that the reward for his long hours is that he is waited on at home. Cheryl doesn't see it quite that way, though. She works full-time off the farm at a small agricultural marketing firm, ten hours a day including the hour-long commute each way. Indeed, it's Cheryl's income that carries the family, "because most of my money gets stuck back into the farm," Floyd had told me earlier during our interview. The family's health insurance also comes through Cheryl's off-farm job. But when evening falls and the weekend comes, Floyd is usually still at it, out on a floodlight-equipped tractor, or in the home office at the computer. Cheryl is usually still at it too, cooking, cleaning, and focusing on the children, and no one waits on her. The family does go out to eat a lot, though, giving her some break. "From Friday to Sunday, I don't really cook," Cheryl had told me earlier when she joined the interview briefly. But then, neither does Floyd, ever.

And when Floyd started extolling the importance of attending the children's various sports events and school performances, and talking about how farming gives him a flexible enough schedule to do that, Cheryl snapped. Their son Tynan has a regular Monday night baseball game, and Cheryl took the opportunity of Floyd's statements to remind him about next week's game. Floyd nodded, then said to Sue and me, rather pointedly, we felt, "You've got to watch Monday nights."

"Well, you almost never go!" Cheryl said, with considerable heat.

It seemed to be one of those times when a partner in a couple uses the presence of outsiders to make a sharp comment, knowing that the enforced politeness of the occasion will prevent an argument from developing—at least until later.

Disputes about time and the sharing of work are a great source of tension among the farm couples we interviewed, and caustic feelings often crept in when this topic came up. After our dinner with Floyd and Cheryl, Sue dropped me off at the farm of Troy and Joyce, a couple with four young children. Troy and Joyce run a joint operation with Troy's brother, together working some nine hundred acres of crops—about half the acreage per family that Floyd and Cheryl work—plus raising cattle and hogs. They farm conventionally, that is, with plenty of machines and chemicals and antibiotics, selling everything through conventional marketing arrangements. They do rotate their crops some. They do consider the manure from their livestock a resource for the farm, and not just a nuisance to be gotten rid of. But they make no special efforts at soil conservation and they "broadcast" their herbicides—that is, spray them over every inch of the field, instead of "banding" them onto the rows and cultivating in between them. The interview started at an awkward time, 8:00 P.M., right in the middle of the children's bedtime routine, and Joyce spent about half of the interview settling the older children in for the night and attending to their six-month-old, who was awake and active throughout our long conversation. Troy looked after the baby a few times, but almost all the child care fell to Joyce, and I think she resented it. Joyce seemed especially interested in the interview when I arrived, but she had to keep getting up and going out of the room to mind the kids.

It was a difficult interview in many other ways. Troy and Joyce are not as well established in farming as Floyd and Cheryl are, and there was a persistent darkness to much of their description of their current circumstances. The interview didn't start that way. Troy in particular was upbeat at first, and seemed eager to make sure that the local farming community came off well. He spoke of the financial pressures in farming as being "just like in any business. It just boils down to making the right decisions." Our conversation turned to the high land rents farmers were paying in the area, and I said, unwisely, "So I gather there's a fair bit of competition."

This was just the kind of comment that books on interviewing warn interviewers against, as it forces a certain evaluation of a topic. "Yes," Troy replied at first, perhaps not wanting to contradict me, especially so early in the interview, as the books warn will happen. "But I'm sure that's anywhere, I'd think," he went on. "There'll be times, I suppose, but I think it's pretty good around here as far as [competition goes]."

Joyce didn't feel the same need to pitch the county, though, and she added, "Well, you know what I've heard is that as soon as somebody's coughing, somebody's saying, 'Oh by the way, if you get sick, you know, I'll rent your land.' That's the comment I've heard."

"I see," I said, trying to be more neutral now.

"You know as soon as somebody is even sick," Joyce continued. "They're not even near their death bed, and people are already thinking, 'Oh, who's going to rent their land?' People I heard were talking to this farmer who was dying of cancer. Before he even died, people were asking him about his land. That's pretty tacky. He's not even dead yet!"

This wasn't the kind of thing Troy wanted me to be told, though, and he put a brighter face on things, at least with regard to their own community. "But as far as the neighborhood goes, we all get along. Everybody's pretty well been farming their pieces for quite some time. . . . Everybody gets along pretty good."

As the interview progressed, Troy and Joyce continued to present two quite different impressions of farm life, and Joyce didn't shy away from making those differences plain. Some of it was due to her different upbringing, she explained, as she was raised in town instead of on a farm. In fact, they live on the very farm where Troy was raised.

"So it's a whole new experience," Joyce said. "I mean, I knew what went on the farms, and I saw animals." Although she wasn't raised on a farm, she did grow up in a small rural town. "Not that part. But just the whole different idea of the lifestyle. We ate at six o'clock and we had regular hours. And my dad was—" My dad was "there," I think she was about to say. She cut herself off, and said instead, "You just are used to a very typical life [in town]. And when we moved out here, nothing is ever definite because something can go wrong. Cattle can get out, and there's work to be done no matter."

"The weather," Troy inserted.

"Yeah, the weather," Joyce continued. "And there's work every minute of every day. Whether or not you choose to. So that was an adjustment. He thinks it's very normal because he grew up that way. That's the way he perceives family life to be. I had to make a big adjustment in how I thought family life was. . . . It's the hours. I guess that was the thing I wasn't prepared for, the hours. And the dollars involved."

"You mean the amount of money it takes to keep the farm going?" I asked.

"Yeah, you know, how many bills they pay," Joyce replied, using "they" to seemingly distance herself from the world of farmers. "Their bills in a month's time was what I made in a whole year teaching school. They write out that many bills

Razing of the Gunderson family home place, a 1907 Victorian farmhouse, in Pocahontas County, 1990.

every month." Joyce had taught school before she and Troy had young children at home. "So I had to get used to that."

But it became increasingly clear as the interview went on that Joyce has not in fact gotten used to the uncertainty and the time demands and financial pressures of her husband's work, or to the way he perceives family life. I had the sense that Joyce was talking as much to Troy as she was to me, and I think from Troy's silence during most of this that he thought so, too. Indeed, his sole comment, the one about the weather, may well have been defensive, an attempt to attribute his workload and the inconsistency of his presence in the home to nature and the nature of his job, not to any choice of his.

That's not how Joyce saw the situation, though. She headed upstairs to read a story to the older children, taking the baby with her, leaving Troy and me in the kitchen. We talked about his approach to farming and marketing for half an hour or so—the kind of topics that Sue, Greg, Donna, and I often found ourselves talking about with farm men—until Joyce rejoined us. I wanted to change the subject, and Joyce's return seemed like a good opportunity.

"Joyce, this is something for you too," I said as she sat down in one of the kitchen chairs, next to Troy and across the kitchen table from me, and placed the baby in the playpen. "We've talked about the way your farm is. What would you say your ideal farm would be? If you could farm anywhere, in any way?" They hesitated, so I followed up with what was probably an overly personal question. "I mean, I'm guessing you're in farming—tell me if I'm right—because you want to be, right?"

"Uh huh," Troy said immediately.

But simultaneously Joyce said, with a questioning tone and the kind of smile that is meant to soften words with sharp edges, "Or if you marry a person who is?"

I was startled, and replied, flat-footedly, "Is that right? What's that?" I wasn't sure I'd heard her correctly, given their over-talking.

"Or if you marry a person who is!" Joyce repeated clearly.

It was an awkward moment, and it left me fumbling to keep the interview on its keel. "Uh-huh" was about all I could manage to say. Troy turned his chair and started playing with the baby, disengaging from the interview, I thought.

Joyce appeared to think so too, and she spoke again, shifting the topic back to the structure of the farm operation, perhaps trying to draw him back in. "I think when you own all your land," she said, "and you don't have a lot of livestock, and maybe a smaller amount of [debt]—"

That worked. Troy spoke up again. "Yeah, I guess the interest. You know, I just hate to pay all that cotton-picking interest."

But this kind of topic silenced Joyce, after her initial effort to turn the conversation to it. So after a few minutes, I returned to the question of the ideal farm, and asked, "Joyce, would you ideally be living on a farm, would you say, or was that—"

"No, I wouldn't. No," she replied. "You know, I love my husband, and therefore I'm going where he has a job. He likes it, which is fine. [But] no, I wouldn't. Who would ask for a person that worked that many hours, and was never home? I mean, nobody would. There are good parts to it. He's home at noon, you know. He can stop in for five minutes in the entryway, periodically throughout the day. That you wouldn't get with a steady job. No, it's, I don't know, a lot of work."

Joyce paused to consider where her words had led her. She tilted her head in a contemplative way. "But there's opportunities that you don't get with a regular job, too. You get some more flexible schedule. Not that he takes a lot of time off. But if he absolutely has to have some time off when going to the doctor, you know you have a flexible schedule there. There are some advantages."

The baby started to cry, and Joyce broke off to pick him up. When he had settled quietly into her lap, she continued, "I don't know, I don't want to sound pessimistic about it. It's not a terrible way of life. But it's, you know, you can certainly—" She searched for words.

And here Troy finished her sentence for her, in quiet agreement, "There's definitely an easier way to live."

Yet, in spite of this moment of agreement, I think Troy felt under attack virtually every time Joyce spoke. At any rate, he lashed out a bit at the end of the interview, about half an hour later. I had asked that standard interview-closing question again.

"Well, usually about the last question I ask folks is, are there any questions that you think I should have asked that I didn't? Did I miss something about your views of farming or farm life? Something you wanted to add? Or was there maybe a place our conversation was going and then we shifted off to something else?"

There was silence at first, and then Troy said, with a low voice and a kind of growling laugh, "Some more deep secrets of yours that you want—"

He didn't finish the sentence, leaving it as a sort of a tease, maintaining the potential cover that it was just a joke—what Erving Goffman once described as "out of frame" activity.[4] There was nothing out of frame about it, though, and both Joyce and I knew it instantly. Rather, there was another rent in the fabric of the situation, much like the one Joyce made with her remark about marrying a farmer. She and I both forced out a laugh, mutually and independently deciding that from our positions in the conversation it would be best to go along with the polite fiction that Troy's comment was a joke, and thus safely out of frame.

But Joyce was hurt. She cut her laugh off, and so did I an instant later. Her face reddened. She turned her attention to the baby again, trying to regain her composure. There was a long pause where neither of them said anything. I too was speechless. The rent in the fabric remained. I later timed the silence from my tape of the interview, and it was nearly twenty seconds—a gap that conversational analysts regard as exceptionally long in the compressed dynamics of live talk.[5]

Finally, Troy spoke, perhaps satisfied that his point had been made, perhaps hoping to maintain his control of the situation, perhaps out of a sense that only he could repair the fabric of politeness, perhaps all of these. "Just get people to eat more beef, I guess," he said, smiling.

This was a better joke. I laughed loud and long, and Troy joined in. But Joyce did not.

Getting people to eat more beef is not a likely solution to the farmer's problem and its consequences for the treadmills of farm family life, however. People able to buy more beef probably already buy as much as they want—the problem of "inelasticity" again. But there may be something to selling a different kind of beef, a different kind of corn, or a different kind of soybeans that people do want more of. There may be something to setting up marketing arrangements that allow farmers to retain more of the profit margin for themselves. There may be something to switching to technologies that may not increase your output, and may even decrease it somewhat, but that cost so much less that your margin goes up considerably. There may be something to farming as a whole family, where there is a farm husband as much as there is a farm wife, where both partners in the couple are "farmers" even if they are doing different things on the farm, and where the children are considered more than a source of farm labor or an impediment to farm labor but rather farm learners, along with their parents.

Donna interviewed one Powell County couple who are attempting to do all of these things, and with considerable success, as far as we could tell. Things aren't perfect for Brad and Jean. They have their struggles, both within the family and the farm and with the world outside. But they showed a level of comfort with what they are doing that was, to our research team, a welcome relief after the tensions we encountered on farms like those of Floyd and Cheryl and Joyce and Troy.

Brad and Jean's four-hundred-acre farm looks different the minute you see it, as you drive up "the gravel," what local people call Iowa's unpaved country roads. One of the first things that strikes the eye is that there are livestock outside on pasture, where you can see them. So many of Iowa's farm animals are now raised in confinement that sometimes one can drive for miles without seeing livestock, despite the fact that Iowa leads the country in pork production and is in the top ten in both beef and poultry production.[6] But Brad and Jean use a method called "rotational grazing" or "managed grazing" for their livestock, and the animals are usually outside, giving a feeling of life to the landscape.

Another technical aside: Managed grazing is a relatively new idea in Iowa and the Midwest. It was virtually unheard of in the region as recently as 1990. The idea is to bring the animals to their food, and to gain the assistance of the animals in producing that food. In a confinement situation, you have to bring the food to the animals. And even when livestock are given a chance to graze, typically in the Midwest they get most of their feed from bales of hay and loads of grain brought out to the pasture by the farmer, as the grass in the pasture regrows too slowly to account for much of the animals' diet. But in managed grazing, a pasture is divided by moveable electric fences into small paddocks of a few acres or so where the animals remain for only a day or two or three, instead of the weeks (or even more) that are standard in an "unmanaged" pasture. The animals graze intensively in the paddock and then move on, leaving their manure to fertilize the regrowth of the just-munched grass. Also, as any lawn owner knows, grass regrows slowly if it is cut too short. If there is little leaf area left to get a recharge from the sun, it takes grass plants longer to put out new growth. And if the grass gets long, it begins to put its energy into seed production, slowing down the production of leaves once again. In managed grazing, the idea is to use the constant rotation of animals to keep the grass at that mid-length where leaf growth is the fastest. Managed grazing also promotes even grazing, which both increases grass productivity and encourages the cows not to leave weeds uneaten to grow and spread. It was a controversial idea among U.S. agricultural scientists at first. It was farmers who brought the idea to the region, based on practices that had been in wide use for several decades in New Zealand and elsewhere.[7] But there is now little doubt, even among the scientists, that it works, sometimes as much as doubling grass production per acre.

Brad and Jean were sold on it long before U.S. agricultural scientists caught on, and on a number of other unusual production and marketing practices as well. At one time Brad and Jean's farm was much like any other in their neighborhood, emphasizing high yields of corn and soybeans by whatever means possible—chemical, mechanical, financial. But for the past fifteen years they have been farming with increasingly less recourse to insecticides, herbicides, "big iron" (as farmers sometimes call the massive machinery of industrial agriculture), and big loans. They now farm entirely organically, and they sell much of their beef through a special farmer-run distribution network that markets directly to consumers who want chemical-free meat raised with methods that are kind to the environment and kind to the animals. It's been a big change, and it has taken them time, but they've done it.

Their home life is different too, as Donna learned during her interview with Jean. They spoke in the kitchen, where we usually conducted our interviews. Donna interviewed Jean separately from Brad. He came into the kitchen briefly a couple of times, on his way back and forth to their farm office, which is in the house. But he pretty much cleared out, a situation we were not always able to achieve in our interviews. Greg and I in particular often felt resistance to our interviewing farm women alone, or even jointly with their husbands, although Donna and Sue had better luck. A number of families seemed to regard farming as the man's exclusive domain, and there may well have been more insidious and extreme manifestations of masculine control at work in some of the refusals and evasions we encountered.

Relations seemed far more equitable between Brad and Jean. Brad takes the initiative on farm decisions, and the responsibility for the house is Jean's, in the traditional arrangement. But Jean also does some of the farm work, and Brad does some of the housework—not fifty-fifty, in either case, but there is more sharing of these duties than we saw on most farms. Moreover, Jean and Brad constantly consult each other, and their arrangement emphasizes the skills of each partner, as Jean described it to Donna.

"[Most of the time] he comes up with the idea himself," she said, discussing decision making with the farm work, "and then says, 'What do you think about this idea?' And then we mull it out. Two-thirds of the time he'll come up with an idea that is about two-thirds of the way decided and planned and spelled out and then says, 'What do you think about this idea that I have full control over?' It's not like it's negative. It is something that he is skilled at. But we don't do anything big without both of us giving a stamp of approval."

This arrangement could not be considered a paragon of equality, and Jean acknowledged as much with her observation that Brad ultimately has full control over these decisions. But she doesn't feel that this is an unjust arrangement. She feels it is a sign of Brad's respect that, even though he does have full control over these decisions, he nevertheless consults her. Besides, he does generally have a better technical knowledge of farm production issues. Like Joyce, Jean does not come from a farm background. Jean feels that she is still trying to catch up with her husband's knowledge base, even though she frequently participates in the farm work. In contrast, for Joyce, farming is her husband's job. She lives on the farm because that's where her husband's work is; farming is not her work. But Jean sees farming as her work too, although she regards her husband as quite a bit more knowledgeable about it.

"My goal used to be that if Brad died tomorrow, would I be able to take over the farm. Would I know enough about it?" she was explaining to Donna as Brad came in from the farmyard.

"I'm leaving in a second!" he said, passing quickly through.

"No mention of a wonderful husband yet," she teased him as he went by.

Donna laughed, and after the door closed she asked, "So, how do you feel now? Say that you would have the responsibility of raising the family on your own. Do you think that you could hold on to the farm?"

"Oh, yes. Management-wise, yes. Labor-wise, no. But management, I feel confident that give or take a few mistakes, I could do it. I'm sure it would be an incredible adjustment because, I mean, our type of farming is labor intensive. But I have the confidence that if that unfortunate situation ever arose, that I would be able to meet it—to say 'yes, I can do this.'"

"Then, what about housework?" Donna asked.

"I started work off the farm a couple of years [ago]—this is my second year—and that helped me realize that adjustments at home [were needed]. Housework? Does it get done like it used to? I don't know. We don't have maid service—which I'm surprised at the amount of people who do work off the farm around here have."

"Are you talking about farm couples?"

"Yeah. That I was surprised at. I wouldn't want it myself. We work together on it." She paused to consider the matter more closely. "Well, not together, but—" She paused again.

Donna pushed the point a bit further, asking, "But who has the main responsibility of it?" Brad and Jean have two sons, and Donna was a little doubtful about how "together" Brad and Jean's family is when it comes to housework, given current social conventions on most Iowa farms—or indeed in most households in the United States.[8] This was also a topic of personal interest to Donna, as she too, like Jean, is the only female on her farm.

"Me," Jean said firmly.

"And then, being you have two sons and no daughters, what are you teaching them about housework and division of labor?" It was a challenging question, but the interview was going well and Donna decided to risk it. Jean took it in stride.

"Hopefully they chip in here," she answered. "They are responsible for their rooms, of course. My goal this summer is to work at teaching them laundry and teaching them cooking. But as far as what have we done so far, you know, the rats right now are winning the rat race in this household. It's like, 'where do we go next?' 'What do we do tonight?' 'What do we have to do here?' So a lot of it doesn't get done. And, to me, it's something I don't feel comfortable with."

Brad and Jean's family has not escaped the treadmill. There's a "rat race" right in their own home, as Jean describes it. Their household remains subject to the treadmill's speed-up of time and of money, increasing the family's need for both. Their home is no haven from these workaday, workanight pressures. Part of the problem is the difficulty they face, as a farm family, in keeping work and home integrated (as it must be, for time management's sake) and segmented (as it must also be, for sanity's sake) at the same time. Jean explained the conflict well, comparing her own upbringing in town with life in her family today. Donna had asked, "Do you consider your home separate from the farm?"

"I wish we could more," replied Jean. "No. I consider it part of the farm."

This too was an issue of personal interest for Donna, who knows the conflict well. Donna had a sense through much of this interview that she was interviewing herself, given the similarities between Jean's farm and family life and her own. Jean seemed to recognize this too.

"It's like for any farm couple, yourself included," Jean observed. "When I grew up, and my Dad had several days off from the post office, he'd come home and he'd putz around the house, unless we went somewhere as a family. Here you don't have that luxury. You don't just say, 'I'm taking two days off,' and come in. You know, it's always there. We each have offices in the home, but there's no escaping. If you need to have a break you have to leave, it seems. You can't just take one in the house, comfortably."

The farm is always there. You have to leave it to get a break. But there's no escaping. Although Jean and Brad haven't escaped, they have been able to move over to the side of the mill where the tread is slower, where they can more easily step off for a moment's reflection about the kinds of demands that really matter. For example, Jean may want a cleaner house, but she has decided that fundamentally it is really not worth the struggle to maintain it. She knows that the image of the perfect household where the super-woman breezily manages a full-time job in the paid workforce and a full-time job in the unpaid workforce at home is just that—an image, not a reality.

"I'm not meticulous. There's a point where I think you win the battle and lose the war. . . . I fully respect the person that can have full-time off-farm employment and come home to a clean house—a clean home, a happy home, a stable home, a home that's not chaotic, if you have young children or kids. Period. I admire them. But my guess is, where are they? Because it has been, you know, hard for us to keep up with things with my job. Where the kids have had to make adjustments and that. I admire people that can do it—do it successfully—but I really question how successful they are at it. And I really question how much extra help they receive—whether it's, you know, nannies for the kids, or babysitters, or other help."

"So what would you say the biggest barrier is to running the home the way you would like it to be?" Donna asked Jean.

"Dirt!" she burst out, and they both laughed. It is a constant struggle to keep the mud and manure of any farm outside, or at least in the "mud room" that most Iowa farm households maintain just inside the back door.

"Running the home?" Jean continued, with a more serious tone. "Sometimes long hours, but not really. Okay? Because that's something we're comfortable with. I am, at least. When the children were smaller it was a more of a concern for me. Like, to use the old adage, 'guess who's coming for dinner tonight, kids? Daddy's going to be home!' When he would be working late hours or something. Now that the kids are older and the kids are part of [the farm operation], helping with their chores and sometimes with their dad's, I've developed a little more flexibility. Years ago it may have been an issue. Either that, or I've mellowed out a little more."

The treadmill of farm life no doubt does accelerate fastest on a young family, such as Troy and Joyce's, with young kids at home. Jean makes this point well, and is pleased that her children are now old enough to spend time with their father through farm work. Also, a young farm family will probably experience more acutely the problems of trying to establish a good track record with the bank and of getting ahead on interest payments enough to hold on to a little bit more of the massive flow of capital on a farm. But Brad still puts in a huge work week, and Jean's new job has placed additional time pressures on the family. They are still very busy people. Although young children and a less financially established farm do place additional pressures on young farm families, there is more behind what has mellowed out family life for Brad and Jean than simply reaching a different stage in their family's life course.

That more may well be the successful cultivation of sustainability—sustainability in their farm, their community, their family, and in their, and our, environment—a sustainability that democratizes them all. Given the uncertainties, the losses, the conflicts, and the struggles of conventional approaches to dealing with what I have been calling the farmer's problem, why haven't more Iowa farmers taken a path similar to Jean and Brad's? To answer that question, we must look beyond the big iron of agriculture's structure.

Intermezzo

At this point, it is perhaps time to say a few more words about the methods of this book. I say "of this book" rather than "of this study" because I mean both the methods of gathering the information I report here, as well as the methods of reporting it. There is as much methodology to writing as to anything else social scientists do.

An aspect of my writing methodology that I imagine is quite clear by now is that I am using a far more novelistic approach than is typical of most sociological research, even qualitative research such as we conducted for this book. By "novelistic" I mean narrative techniques like the little descriptions of setting and mood that I often include with the quotations. The way I focus on just a few individuals in each chapter, rather than backing up each point of my argument with a series of brief quotations from many individuals. The write-up of these long extracts from our interviews and field notes as scenes with narrative movement. The rapid cutting from scene to scene, as if they were simultaneous subplots in a larger story. My own presence, and that of my colleagues, Donna, Greg, and Sue, as full characters in this larger story.

These techniques make the book more compelling reading, I hope, building readers' enthusiasm for continuing on through its many pages. But there is some considerable rhetorical hazard here too, at least for a work of social science. A book may seem less factual the more it appears to draw on the narrative strategies of fiction. Without getting too much into the philosophical imbroglio about what "factual" means, let me just say that many readers might imagine that my writing less faithfully represents the circumstances of the research the more it adopts the manners of the novelist. If it reads like fiction, perhaps it may give readers an uneasy sense that it indeed is fiction.

Given this hazard, I ought to have a better reason for novelistic writing than only the hope of being more entertaining. And I believe I do. Although I will be happy if the book reads better as a result, my main goal in using the techniques of fiction writing is, perhaps paradoxically, to be more factual—that is, to better represent the evidence I report here. To render a quotation or other social action stripped of the context in which it was given seems to me to be stripping it of most of its sociological content. Context is content. The great weakness of quantitative sociological research is that its strength is based on its efficiency in decontextualizing evidence, kidnapping social practice in all its wondrous complexity and hiding it away in tables of numbers that present verity as the reduction of variety.[1] The practice in much qualitative writing of giving a long series of short quotations from many people to document a point is to participate, albeit to a lesser degree, in this same methodological mass kidnapping. My approach, by contrast, is to provide substantial context for the evidence I present, using the techniques novelists have honed over the centuries.

I don't mean here to argue for a correspondence theory of truth. When I say that my goal is "better" representation I do not mean to approach that betterness through making my descriptions identical with the social practices they portray. The only identical representation of an aspect of reality is that aspect of reality itself, including its contextualization within the universe that gave it rise. It is a tall order, taller than we can imagine, to place the entire universe into the pages of a book. All representations of social practice are just that: representations, not the real article, as the postmodern critique of modernism's positivist pretensions has made clear. Moreover, I must choose to represent specific aspects of any social practice that I describe within these pages, not the whole thing.

Which is just fine, great, in fact, given the impossibility (and I would also say the superfluousness) of identicality—as long as I choose the aspects that are relevant to the analytic task at hand. My success in choosing the right aspects, and in communicating them to readers, is what I mean by "better" representation. Meaning, as pragmatists have long argued, must always be connected to purpose. (This is not, however, an unproblematic view, as I take up in the fourth intermezzo.) And among my analytic purposes is taking into account the role that I and my colleagues play in constituting the evidence we report here, as well as the particular effects on evidence that research of the sort we conducted may have. The words of Iowa farmers I include are words that were said to us in the settings, moments, and moods in which we and these farmers found ourselves brought together. I have done my best to take all this into account in interpreting the significance of their words. I may not always have done so successfully, though, and I can better alert readers to that possibility and better provide them with the information to make their own judgments for their own purposes if, within the limits of readability and dialogic stamina, I at least make an effort at contextualizing the contents of this book.

In any event, lest some readers find fault with this approach, I hope I have alerted them to it, so that they may more easily take it into account in evaluating the arguments I make in the book.

Let me also point out some real fictions that I have found necessary to include, in order to ensure the confidentiality of those whose words appear here. With only four exceptions (in addition to me and my colleagues) everyone who speaks in these pages has been provided with a pseudonym, first name only. For these fifty-three pseudonymous individuals, I have also altered a few details of their lives and contexts to provide them with plausible deniability, should anyone try to identify them. (Depending on how one counts things, about another twenty to twenty-five Iowa farm people make some kind of named appearance in the book, usually pseudonymous, and usually in association with those I quote.) In a work of social science, such deliberate fabrication is best avoided if possible, as Mitch Duneier almost always managed in his recent pathbreaking ethnography of sidewalk vendors in New York's Greenwich Village.[2] But without this protection, few of the participants in our study would have agreed to be part of it. Iowa farmers are a private people. In most cases, changing these details and not revealing names has little bearing on the arguments I present. But the words of four individuals—Dick Thompson is one who has appeared already—could not be detached from their names without compromising my ability to describe the development of sustainable agriculture in Iowa and the place of PFI in that development, and I refer to them by their first and last names. In these cases, the individuals concerned read over the entire manuscript and approved of my printing their quotations, in a couple of instances pointing out a few errors I agreed I had made but requesting no deletions of any of their words. I also made the manuscript available to any other member of PFI, and a handful took me up on the offer. I was saved from several more errors through their interventions.

I would also like to caution readers against getting out a map of Iowa and trying to locate Powell County. You won't find it. PFI farmers are few enough on the ground that if I reported their actual counties I would have had to describe their individual farms in only the most general terms, lest their confidentiality be compromised among those knowledgeable about the Iowa farming community, and particularly the sustainable farming community in the state. Non-PFI farmers would probably not be so identifiable, but I felt it would be awkward to treat them and the PFI farmers differently. My colleagues and I concentrated our research in several Iowa counties, and Powell County represents something of an amalgam of these.

And now, let's head back there.

PART II

THE CULTURE OF CULTIVATION

4. Farming the Self

I was driving a bit too fast for a country lane, especially in a car with "Iowa State University" painted on the door. (It's bad for public relations, and relations with one's department head, to be caught speeding in a state university vehicle.) But I had been invited to observe Rob, one of the biggest farmers in Powell County, at work with his 360-horsepower John Deere 9300 tractor and his sixty-foot-wide, thirty-six-row planter. I hadn't ever spoken to Rob. The appointment had been set up by a third party. So I didn't want to be late. It was a warm, dry spring day, perfect for planting, and the dust spiraled out behind my wheels as I swayed along the rutted and washboarded gravel road. I was also trying to keep half an eye on the local plat map, an essential resource for keeping track of the constantly shifting tangle of ownership and rental arrangements on Iowa grain land. The third party who made the appointment had scribbled in a little "x" on the map to show me where I might find Rob and his planting crew.

If you farm like Rob, it does indeed take a crew to plant. Rob owns about two thousand acres and rents another four thousand scattered across a five-county region, centered on Powell County. This makes his operation some seventeen times larger than the Iowa average and six times larger than the magic thousand. Farming more acres doesn't increase the length of good planting weather, though, and the only way Rob can work six thousand acres, spread over a seventy-mile-wide area, is with some of the biggest farm machinery on the planet. He also needs a support team of at least three: one to drive the field cultivator ahead of the planter, one to drive the grain truck for reloading the seed boxes on the planter's back, and one to drive the pickup to town for spare parts when the planter (or anything else) breaks down.

Which is exactly what had just happened as I pulled up to the hilly section where Rob and his crew were working. A huge red Case tractor was tearing up and down the slopes, swiping a field cultivator across the face of the hills. I flagged the tractor down. A man in his early twenties, stripped to the waist, his bodybuilder's chest shiny and smooth as a can of beer, jumped down from the cab. "Are you Rob?" I called to him from the edge of the field.

"No," he said shortly, half ignoring me, as if he had stopped the tractor for some reason having nothing to do with my presence.

"Well, I'm down from Iowa State. I'm supposed to ride around with Rob this afternoon."

"He ain't gonna have time to answer questions," he said, knocking a clod of sod out from where it had hung up in the machinery, and still not looking at me. "He's just broke down in the next field."

But he sent me along anyway. I hopped across a narrow stream to the field where the big green John Deere sat crouched, the planter on the back folded up on one side and half folded up on the other, like an enormous dragonfly with a broken wing. A grain truck and a pickup sat out there with it, and two older men were standing around. As I walked up, a slim, tight-looking man in his mid-thirties climbed out of the Deere—jeans, cowboy boots, a western shirt, a cap I didn't take note of, and a pair of pliers in a holster on his wide leather belt. This was clearly Rob. He was just slipping his cell phone into a holster on the other side as I came up and started to introduce myself.

He cut me off with a harried glare. "I'm broke down," he said, dispensing with any formalities. "Might be an hour or two before I'm up and running. So I don't know if you want to stick around." In other words, don't.

Rob had good reason to be unwelcoming. It's a huge gamble, telling his bankers and landlords that he can plant six thousand acres in eight weeks—eight weeks that are among Iowa's rainiest. And this spring was particularly wet. Everyone was behind. Rob didn't have time to break down, and he was plainly reconsidering the wisdom of allowing himself to be talked into having me come out to ride around with him in his "big green machine," as farmers call the top-of-the-line John Deeres.

The polite thing, clearly, would have been for me to leave. But Rob had already put me out of his mind, and had gone back to thinking and fuming. And I really wanted to stay. I was as interested in the fixing of a broken-down 9300 as I was in the running of a fully operational one.

"I hate this," Rob spat out to no one in particular, as he eyed the lame planter wing, askew in the air. "I just can't stand it!"

Then Rob saw what the problem was. He had been folding up the planter, in preparation for moving it to another field, when a hydraulic hose came loose somehow, got caught in the framework as it folded, and was severed. With the loss of hydraulic pressure, the wing couldn't be folded or unfolded any further. Rob whipped off the hose with his pliers and tossed it to Darrell, one of the men standing by—Rob's father-in-law, as I later learned. Darrell tossed the hose into the back of the pickup and roared off. Rob climbed back into the cab of the 9300, I guess to get some peace while he phoned the parts store to tell them Darrell was coming. It was 3:50 Saturday afternoon. The store closed at four on Saturdays. No way was Darrell going to make it by four.

Early winter application of anhydrous ammonia fertilizer to a tilled field, Franklin County, 2000.

In Rob's favor is the fact that the bankers and landlords have taken just as big a gamble as he has. If Rob goes bankrupt—if he can't keep his equipment and his crew running, if he can't get a crop out of the ground while keeping his costs down, if he can't get a good price for the 1 million or so bushels of grain he grows every year—it's bad for everyone Rob owes money to. He has to pay mortgage payments on the land he owns, and rent on the land he doesn't, at least $100 an acre in either case. That's $600,000 a year right there, and probably quite a bit more. Then there's his $175,000 tractor and his $200,000 combine. Rob gets a new one of each every year because of the wear he puts on them.

Then there's $25 an acre for seed, $40 an acre for pesticides, $50 an acre for fertilizer, $35 an acre for fuel and machinery repairs to put the crop in and to harvest it, plus another $20 or so per acre for crop insurance and other miscellaneous items of the sort that always seem to beset any complex endeavor. There are also wages for Curt, the guy driving the cultivator with the Case tractor when I first arrived, and wages for Vern, who drives the grain truck. (Darrell works for free.) If you work it all through, it comes out to about $375 an acre on his corn land and $275 an acre on his soybean land, assuming Rob is able to meet the "estimated costs of crop production" figures that Iowa State University economists calculate each year. That's about $2.75 for every bushel of corn and $6 for every bushel of beans, if Rob produces average yields—about $2 million in all.[1]

In other words, a lot of people's fortunes funnel at least in part through Rob's.

They don't want him to fail. But many farmers do, every year, because it has been a long while since Iowa's farmers got much more than $2.75 a bushel for their corn or $6 a bushel for their beans, even with crop subsidies.[2]

Big dollars, big machines, big acreages, and smaller and smaller numbers of farmers. It may not be the best way to grow corn, for it turns rural Iowa into a high-stakes casino with few winners and lots of losers. The losers include individual farmers and their families, and also communities and the land. But that hardly mattered that big green afternoon, because the real crop Rob is trying to grow out there isn't corn. It's his self.

Rob's self was feeling better as he climbed back down from the cab of the 9300. The parts store had agreed to stay open until Darrell got there. The parts store doesn't want Rob to fail either. With fewer farmers out there, the parts store is increasingly dependent on each one that remains, and particularly the farmers with the big acreage and big tractors.

I couldn't help thinking as I watched Rob climb back down that maybe he liked having all these people dependent on him. Sure, it was great on purely instrumental grounds to be able to get the part he needed. But there was also grandeur in having others orient their lives around your needs.

Even if I was right in this suspicion, Rob didn't take much time to indulge himself with a warm bask in the glow of power. He was right back at it the instant he stepped off the ladder, moving at a speed and intensity I eventually found tiring to be so close to. He set about refilling the seed boxes on the planter with more bean seed (it was actually soybeans that Rob was planting that afternoon, not corn), with the help of Vern, who maneuvered the grain truck into position for easier refilling. They had a long hose and a vacuum pump connected to the grain truck, and Vern used it to shoot the bean seeds into the boxes as Rob opened their lids one by one. I tried to help out by snapping their lids back down afterward, but Rob didn't like the way I was snapping them down.

"We'll take care of that, thank you very much," he barked at me, giving all the lids I'd done a big whack with his hand. I was being too forward, to be sure. Still, I mused, Rob is evidently the kind of man who likes control, not help.

It was an efficient use of time to refill the seed boxes while Darrell went to town for a new hose. But it had a make-work quality to it, and the frustration was plain to see on Rob's face. "Shit, I hate this," he said again.

Then, after a few more boxes had been filled, perhaps figuring it beat make-work, he said to me "So, you're writing a book. What's it going to be about?"

"Farming in the new century," I replied, giving him a version of my standard line, "which is why I'm here." I pointed to the tractor and planter. "This is sure the new century of farming."

"Yeah, I guess it is," he said without enthusiasm.

I dropped that tack. I wanted to know about his environmental practices, and I wondered whether he just uses chemicals to control weeds or whether he also cultivates the crop for weed control, aside from smoothing out the residue from last year's crop before planting, as they were doing today. Cultivating for weed control is a practice advocated by most sustainability enthusiasts. After a bit I asked, "So, do you do much cultivation?"

"Not much, no. Takes too much time."

"Yeah, in this kind of operation," I said, trying to sound at least faintly knowledgeable as well as to sound understanding at the same time.

As Rob and Vern filled the seed boxes, a man in his late sixties, I judged, drove up to the edge of the field in a pickup and walked over. He seemed to be unknown to Rob and Vern. They ignored him.

"Wow, how many rows is this?" he wanted to know. They didn't answer.

"Quite an investment," he tried again. Rob looked up a bit at that one, but continued working.

So the man talked to me. It turned out he's retired and had been fishing in the area, and often hunts pheasant here too, making arrangements with local landowners ahead of time, including the owner of the very field Rob was planting. "I got sixteen birds just in this field," he told me. He was raised in Powell County and now lives in a city a hundred miles away. But he likes to hunt and fish in his home county.

He called over to Rob, "So how do you control the cup grass?"

It was quite out of the blue, but this time Rob answered, mentioning a few herbicides. (I didn't take note of which ones he mentioned, but Arsenal, Authority, Fusilade, Prowl, Pursuit, and Round Up—with their macho and militaristic names of the type herbicide companies often favor for their products—are among those that "take out" wooly cup grass and other grass weeds.)[3]

"It's our biggest problem out here," Rob went on. "But they'll come up with something in the next couple of years that'll take care of it."

The man and Rob then struck up a conversation about GPS—global positioning systems—the satellite and computer technology that some say will one day make it possible to farm without anybody in the tractor at all. It's already possible to use GPS to control planting rates, to keep track of yield and pest damage differences within a field, to regulate fertilizer and pesticide applications accordingly, and to drive a tractor faster, straighter, and easier. A number of farmers in the past few years have been installing "on-board GPS" in their tractors. It's glamorously high-tech, gee-whiz stuff. Rob was planning to get it next year. "But I don't know how good it will work out here, especially in fields like these. It's probably more appropriate for really big flat areas."

They finishing loading the seed boxes. The man left, without a word, and Rob was back up in the tractor making more calls on the cell phone. So I chatted with Vern. I asked him how many horsepower the tractor was.

"I don't know. Have to ask the 'chief,'" he replied in a half-taunting voice, but well out of Rob's hearing.

I soon saw where that taunting tone came from. Rob seemed to pause in his phone calling, and I thought I might get in a few words. I called up to him in the tractor cab—the door was open, so he could hear me—and asked how many people he had working out here, and how many of them it takes to do the planting. I wasn't sure yet that I'd figured out how the operation was organized.

"It's mainly Curt and me." Curt was the bodybuilder driving the cultivator in the next field.

"What about Vern? Does he work full-time?"

"Vern?" Rob positively sneered. "He can't hardly do—" He paused to see if Vern was in earshot and thought better of completing the sentence. "Vern doesn't know all that much," he said in a milder tone.

Just then, Darrell pulled up in the pickup with the new hose and Rob sprang back into action. "Get me some thread tape. I think there's some in my tool box," he shouted at Vern, while he looked over the hose and trimmed it a bit with a knife.

Vern rooted around for a minute in Rob's toolbox, which was mounted on the side of the tractor. He came up to us with the requested tape. Rob glared at him and barked out "sealant" to Darrell instead. Darrell went back to the tool box and returned in a minute with a can of thread sealant.

"See, Vern couldn't find that," Rob said, mainly to me.

"But you told me to look for thread tape!" protested Vern.

"Vern, you've got to think. You've got to expand your mind," Rob said, cuttingly.

I doubt Rob would have said that to Darrell, his father-in-law. Vern was left with little immediate recourse but to continue his half-taunting references to "the chief" and "the boss" when he could to me in an aside, as he did a couple more times.

After another hour, they finally got the planter working again. It was a huge delay for them, and I was worried that Rob would cancel the ride in the big machine, the putative reason for my being there. But to my relief he didn't. I climbed up the ladder into the cab after him. Almost all tractors have only one seat, and this one was no exception. I perched myself on the left arm of the driver's seat, as is customarily done when someone rides along. The manufacturers usually make the left arm extra wide for this purpose, leaving the right side clear for the main controls. It's a bit of a precarious and cramped place to sit, though.

He turned the key. The engine roared.

"Now if I'm in your way, let me know," I said, in a bit of a shout.

"Don't worry," he replied, slamming the door shut on the cab. "You better believe I'll let you know. And fast."

A few weeks later I visited Dale's four-hundred-acre farm. He was haying that day with his teenage son, Jeff. They were laying in some "old-fashioned square bales," as Dale referred to them, contrasting them with the big round bales that only a machine can lift, increasingly common on farms that raise hay. He thought I'd enjoy an afternoon's hard work on the land and a chance to see his farm up close. I thought so too, so I took him up on the offer.

We talked about many things that afternoon as we worked, stacking the bales on the hayrack after we yanked them from the baler with a bale hook. Dale's was an older-model baler that he said his dad bought thirty-five years ago, and the bales didn't pop out automatically. It was pleasantly exhilarating to stack the bales high as we bounced across the field, balancing on the hayrack. It tested the limit of my strength, and I felt it the next day. But it was fun.

It wasn't just fun, though. The whole afternoon was premised on Dale's desire to show me how he does it differently from the "BTOs," as many farmers call them—the Big-Time Operators. Dale's farm is only a few miles from some of Rob's rented land, and Dale was well aware of Rob. When you farm only four hundred acres yourself—a figure that is itself above the state average—it's hard to ignore a near neighbor with six thousand. Rob came up in conversation a couple of times. Dale knew that I had spent some time with him, and he was very curious to know what I thought about him.

"What motivates a guy like that?" Dale asked.

"Well," I answered, "it's a power thing, I think."

Dale nodded his head. "Yeah, power. It's an ego thing."

But to leave it there is too simple, I came to realize. My afternoon with Dale tossing square bales around was an ego thing for him, too. I was there in part to validate Dale's way of doing things. As a professor in the state's college of agriculture—a sociologist, admittedly, but an agriculture professor nonetheless—it was something of a trip for Dale, in the sixties sense of the word, to have me around for the day.

And, as with Rob, running the farm often entails having other people orient their activities around Dale's. That afternoon of haying there were two of us at work with him, Dale's son Jeff and I. We worked late, and when we returned to their house, Dale's wife Yvonne laid out a supper that was carefully planned to be ready soon after we arrived: hamburgers, white bread, chips, pretzels, canned green beans. Yvonne works out of the house too, on-farm but out of the house, mainly with their small goat herd. Yet she sets the rhythm of her day around the preparation of meals for Dale, as far as I could tell, as well as taking care of most other household chores.

My point is not that Dale and Rob have the same amount of power. I don't think they do, not even close. Far more people orbit around Rob than around Dale. There are Rob's full-time employees, his father-in-law, and his service providers, like the parts store. (Dale doesn't have any employees, and I suspect the parts store wouldn't be quite so ready to stay open late for him.) There are Rob's creditors and landlords, who also have considerable power over him but must contend with Rob's considerable financial gravity nonetheless. (I suspect creditors and landlords find that Dale exerts quite a bit less fiscal pull.) And there is Rob's own wife, Mindy, who does all the housework and cooking in their family, as Donna found out when she later interviewed them. (But in this, Rob and Dale are much the same. Indeed, this arrangement prevails on almost all Iowa farms.)

Nor is my point that Dale desires, as much as Rob, to have power over others. I don't know either one of them well enough to say that power is more or less important to Dale or to Rob. The existence of power differentials is no sure measure of the degree of one's desire for power.

Nor should we necessarily fault either of them for desiring at least some power in their lives. It would be a sad situation indeed if a person lacked the capacity to influence those around him or her—to gain recognition for his or her legitimate interests and assistance in attaining them. The political issues are how we ought to define legitimate interests and the degree of assistance each of us lends others in attaining their own legitimate interests, not whether we should have such interests and assistance.[4]

I have no doubt, however, that Dale and Rob find some personal foundation in the way their actions in the world—and the actions of others—manifest their capacities to influence others and their understandings of legitimate interests. Which leads to what I really am trying to say: that for them both the real crop out there is the self.

Central to those capacities and understandings for Iowa farmers are the routines and imaginations of gender. Much of what I observed those afternoons with Rob and Dale were the manifestations of the social performance of masculinity.

I say manifestations in part because I like the pun (man-ifestations) but in part to emphasize masculinity's performed quality. Masculinity isn't something that just is; it is something that is done, and redone and redone and redone, often in ways calculated to ensure its visibility. The same can be said of femininity, both in town and in the country. We don't just store gender in our minds and bodies; we restore it, over and over and over again.

Restoring masculinity is just what Rob was doing when he refused my help in filling the seed boxes. It's what he was doing when he put Vern down in front of me and Darrell, and when he selectively ignored the questions and interests of others in the field that afternoon. It's there too in his continual assumption of a hard-driving, take-charge persona.

Not all masculinity is of this tough, cowboy-boots variety, however. We should not always speak of masculinity in the singular, but rather also of masculinities, as the sociologist Robert Connell and others have pointed out.[5] And central to every masculinity is its social constitution. Rob wasn't the only one performing gender out there, nor was he the only one performing *his* gender. The manifestation of Rob's masculinity required the stage and acting company of the planter and the planting crew—a planter to give a story line and a planting crew to demonstrate Rob's take-charge character. Walk-on players like me and the fisherman who stopped by provided additional opportunities for the theater of gender.

Dale was equally engaged in the restoration of masculinity through social performance. My validation of his approach to farming was also a validation of him as a man, for his is a very manly style of farming. He may use old equipment, but he's not afraid of the hard physical work it entails. He takes masculine pride in his ability to throw a bale high. He takes rugged satisfaction in his commitment to farming his own way, eschewing the high-tech path extolled in the farm magazines and at the coffee shop in town. And it helped that I was there to be an audience for it all, as well as a supporting player.

One can take the theater analogy too far, though. For one thing, one has to be careful not to ignore the fact that in the real social play of gender the actors participate in writing their own scripts, and they also try to direct the action. They are not merely playing, but also struggling, struggling to both be and create a character, a self. All the world's a stage, of course, but we should not take this to mean that we are just empty souls reading lines that society has written for us.

But we do take cues from our social surround, and the cues of gender are mighty among them, at least in farming.

One of the cues is the very language of farming and farmer. When farm people talk about farmers, they almost always talk about farmers as male.[6]

Take Marlan and Kelli. They farm about nine hundred acres along with two other couples, Marlan's dad and mom and Marlan's brother and his wife. They all have active roles in the farm operation, but when Marlan and Kelli talk about it, it's the men who do the farming.

Sue interviewed Marlan and Kelli in their home on one of the farms the three couples rent, close to the hog confinement barn that is Marlan's principal responsibility in the family operation. Marlan's parents own some land, but the brothers and their wives are mainly renting. Marlan and Kelli were not always the most forthcoming of interviewees, and Sue was finding that she had to keep returning to the printed list of questions that we usually brought with us to our interviews, hoping that we wouldn't have to use it much to keep the conversation going. Sue flipped through the list, and scanned it for something she thought would help them articulate their understanding of farm life.

"Quality-of-life questions," said Sue, that having caught her eye. "How do you define—and in a way we've touched on this too—a good quality of life for yourself?"

Marlan had been doing most of the talking, as we usually found that the men did when we interviewed couples together. But this was one for Kelli, he seemed to think. Or maybe he just didn't know what to say. He sighed and turned to her, saying, "What do you think?"

It's not an easy question to answer, and quite possibly Kelli also thought it was potentially a sensitive issue between herself and her husband. She hesitated. "I think what we're living, you know, is pretty much—" she began.

"I'm satisfied," broke in Marlan.

"I just wish we had a little more put away, that's our thing," Kelli continued. "I would like to have something to call my own. That's one thing that is big with me."

"Meaning a farm?" asked Sue.

"A home."

"A home, a house," said Sue, trying to stay tracked with her.

Kelli nodded. "You know, this is home, but it would be nice to have something of our own."

"That makes sense," said Sue.

Marlan took an expansive breath. "We're kind of in a position," he started to say. Then he abruptly shifted the topic to a clicking noise that had been in the background during the entire interview. "That's my hog medicator, that clicking all the time."

Maybe Marlan just wasn't interested in the topic, or maybe he thought Kelli was getting too personal. Maybe Kelli was indeed raising a bit of a family sore point, perhaps something Marlan felt some inadequacy about as the economic captain in a traditionally patriarchal family. Whatever the reason, suddenly the conversation was about in-home hog medicators.

"It's very fancy," said Sue. "Sounds like it's doing hard work. So it goes here out to there?" meaning the hog confinement barn just fifty yards from the house.

"Yep. One ounce of medicine goes out for every gallon of water."

Sue, who is knowledgeable about Iowa farming, had never heard of this kind of setup. "Keeps it warm, and easy to monitor. Is that typical to have it in the house?" she asked. "I've never heard of that."

"Wherever the main water system is," Marlan explained. "A lot of guys are doing that. It's all enclosed. It can't be messed [with]. The kids can't get at it. I mean, it's all contained. It's warm here. I just go from the TV to there, and I re-adjust it and go back. You don't have to wrestle waterers around."

A lot of guys are doing that: The language of farmers as men, as guys, slips easily into rural conversation in Iowa. Of course, the word "guys" can be used in a gender-neutral way in the United States, and often is. But the weight of the word is male, and everyone seems to understand it that way in the Iowa farming community when the subject of farmers comes up. Women, too. Later on in the interview, Sue brought up the subject of the tightness of the land market. The conversation was going better now, and Sue had laid the list of questions down.

"Is there enough land?" Sue asked. "This is another thing we've found. Some farmers would say, there just isn't enough land. . . "Sue was probably leading the interview a bit too much here, as we all sometimes did in our field work. It's a hard thing to avoid.

"It's the same thing here," replied Marlan, agreeably enough.

"I wonder if that's everywhere in the community," Sue added, trying to prompt Marlan and Kelli to say more, and on their own terms.

After a moment Marlan did, explaining that some in the community see more of a shortage of farmers than of land. "See, there's a generation of farmers coming of age here for retirement. I've had several guys tell me that they think five or ten years down the road that there won't be enough farmers to farm the land."

"Yeah, I had somebody tell me that this morning, and I was surprised at that," said Sue.

"Yeah, I don't know," Marlan said, incredulously now, perhaps picking up from Sue.

"Yeah, I was skeptical too," Sue agreed.

"Yeah, I'm not that—" Marlan began, reaching for his handkerchief.

Then Kelli took the opportunity to speak. "There is going to be that guy that's going to pick off more than he needs. There's always going to be that."

As she spoke, Marlan blew his nose loudly.

Here Kelli gave a more negative reading of social relations between farmers than Marlan did, as Joyce did in my tense interview with her and Troy in Chapter 3, and as we found that women often did. It's not a matter of a shortage of land or a shortage of farmers, she suggested, but of competitive greed among farmers. But Kelli made her statement in the language of "guy" and "he"—perhaps deliberately, in this case, to place this competitive behavior squarely in the realm of men and masculinity. She may also have been chafing over Marlan's picking off more of the conversation than he needed.

But Kelli presumed, as she adopted the language of "guy" and "he," that Marlan and Sue would know exactly whom she was talking about. In the transcript of this interview, the word "guy" or "guys" appears thirty-three times in a little more than an hour, about once every other minute. In twenty-two of those instances, Marlan or Kelli uses "guy" or "guys" to refer to a farmer, to a group of farmers, or to farmers in general, and Sue uses it that way twice as well. A lot of guys is right.

And wrong, for it excludes women. It is true that on most Iowa farms men put in more hours of labor outside in the fields or in the animal barn and the machine shed. But given the close integration of home and work in farm life, the work that goes on inside the house—the barn for the humans—and off the farm premises is as central to the success of the operation as the areas of farm labor where men clock the most hours. Moreover, on many farms run by couples, women put in a large number of hours in the fields, in the barn, and even in the machine shed—not to mention the farm office, where they frequently put in more time than their husbands and male partners. And, of course, there are farms run entirely or principally by women, although these amount to only a few percent of Iowa farms.[7]

Yet even the women who do put in a lot of hours in the tasks conventionally seen as farm work often have trouble claiming the self of "farmer," particularly if they do that work along with a male partner. I interviewed Shelly and Bert together in their cluttered home on their small hog farm, where they raise a few thousand free-range hogs every year. I had visited their farm a number of times and had gotten to know them fairly well. The interview had a free and open quality that I think we all enjoyed, and started out with a sumptuous lunch prepared by Shelly.

Shelly and Bert are a close couple, and I had already noted that they work together on many tasks in the fields and the barn. Given the masculine language of farming, I was eager to know how they thought about their work together.

"So," I asked, "how many hours would you say that you folks put in a week between the two of you?"

Neither of them responded at first. Hours of work per week didn't seem to be a number they keep track of. I tried to give them a point of comparison, asking "Do you work a forty-hour week?"

"A couple times a week!" said Bert with a rueful laugh. "Yeah. I'm doing twelve and eighteen hours right now through the winter time. We've had some unusual circumstances and it'll be that way again during farrowing. Yeah. The hours are crazy."

"Shelly, how about yourself?"

"Well, does cooking count?" she asked in reply. "Since I do all the cooking and the housework."

"Well, what do you think?" I wanted her to frame the issue as much as possible.

"Well, I—" She seemed reluctant to call her cooking and housework farm work. "For the pig work, I'm probably doing forty or fifty hours a month."

But she seemed to want to excuse this. "I'm basically only out there partly because we can't just afford meat that much," she went on. Commodity prices were quite low at the time of the interview, and she and Bert were feeling a bit pinched. "I'm out there when there needs to be a second person."

"So that means like every other day, kind of thing?"

"It just depends. Like when we sort pigs to sell, I'm out there. . . . When we have the vet out, I'm usually out there. But otherwise I'm doing housework."

She paused. "That counts too," she said finally.

"Right. Right. Right," I said, probably over-eagerly, but trying to be supportive. "I think it does, but I wanted to know what you thought."

"Well, basically Bert does his own laundry, and I do everything else." She gave a bit of a laugh. "He helps clean. I'll give him that."

Bert had been listening intently to the conversation. In response to our questions, couples often brought up issues in the interviews that apparently they hadn't discussed fully among themselves. I think we created some family tension sometimes, but it also was a chance for partners to learn about each other as we learned about them. Communication isn't always perfect with any couple, after all.

In any event, Shelly was saying things Bert hadn't heard before. "Would you like to be out in the fields more?" he asked, seeming a bit surprised.

"Yeah."

"It kind of depends on what's going on, probably."

"Yes."

"This time of the year it can be unpleasant," he went on. "It's, you know, zero degrees. And below, with the wind. But sometimes when the piglets are farrowing, it's really enjoyable."

"Yeah," she said. I wasn't certain whether Bert was talking to Shelly or to me, and I think she wasn't sure either. If his remarks were directed mainly at her, then they had a definite patriarchal air to them, making the claim that the work was hard and unpleasant, and therefore a man's job. If his remarks were directed at me, then they were perhaps intended to be descriptions of Shelly's own attitudes (and still patriarchal in that he was speaking for her). Or perhaps they amounted to a bit of masculine swagger about his ability to endure tough conditions. It was probably all of these. But Shelly wasn't ready to sort out just then which it mostly was, and offered only her brief affirmative statements.

Bert seemed to recognize that he was treading on shaky ground, and kept the conversation going on the fun of farrowing piglets. "It's like Christmas to go out there and see what happened."

"Right," I nodded.

"And look to see what you need to do to prevent trouble."

"Right," I said again. Shelly had fallen silent, leaving me to add in the little affirmatives that support conversational connection. She was thinking. Bert started to say something else, and then Shelly cut in.

"I really enjoy doing the work when we're doing it together," she said, looking right at me, but I'm sure with a wider audience in mind. "When I'm sent out to do it by myself, like put up fence or something like that, it's just work."

This was an entirely different way of thinking about the relative pleasantness or unpleasantness of farm work from what Bert had started to describe. Shelly doesn't mind the cold. She doesn't object to being sent by Bert to do things (or at least didn't object just then). She just wants to be doing things with Bert.

But this more relational understanding of work and farm life was not a map that Bert could easily read. Perhaps he experiences working together as a dependence that undermines masculinity, not a source of human connection. I'm not sure. He certainly didn't pick up on what she said.

"I don't really look at it as hours," he continued, as if she hadn't made a comment. "People want to know how many hours you work. A day is a series of problems to solve. You know what a few of them are going to be, but you don't know what all of them are until you've been out there."

There are two differently gendered selves living and working on Shelly and Bert's farm and in their relationship. These differences don't always line up in a synergistic way. For Bert, work is a rational process of solving problems in difficult conditions, although it can also be fun. Shelly probably wouldn't disagree with that description. But she is very clear that a major part of the fun is doing things with somebody else. Bert probably wouldn't disagree with that either, but either this dimension of fun doesn't come as readily to his mind or it is something he doesn't feel he's supposed to emphasize—at least in the presence of another man. Again, quite possibly all of these.[8]

But of these two selves living and working on their farm, one is clearly the real "farmer" in their minds. Later on in the interview I introduced the question directly. Shelly hadn't been saying much while Bert talked about how their free-range approach differs from a standard hog confinement operation.

"Hey, Shelly."

"Yes."

"Would you consider yourself a farmer?"

"Well, I was raised on farm, and then I got away from it, and now I'm back to being it. Yes and no. It depends on who I'm talking to."

"What do you mean who? Who would you feel more comfortable—?"

Who would you feel more comfortable telling you were a farmer, I was starting to say. ("Whom" is what we should have been saying for something that is now in print.) But she understood before I finished.

"Well, what difference it makes to them."

"Yeah."

"Deep down I consider myself an artist." Shelly is a painter, with a master's in fine arts from a top art college. She's been trying to get a studio set up on the farm but hasn't been finding much time to work on that recently. "However, when I think about it, I probably know as much about farming as I do about producing art. I just right now don't happen to be putting as many hours in as he is."

"Right. Right. Right."

"He's the farmer. When I write down my job description, whenever I have to do it—not job description, but title. You know, like on an IRS form?"

"Right."

"I'm a swine herdsperson."

"But you don't write down farmer," I said, nodding, and noting her careful use of "herdsperson" in place of herdsman. "Right."

"No. Well, now, there. I could get in trouble by writing down farmer. Because if I'm a farmer, I have to be filing whole different kinds of things."

"Different kinds of forms. Right." I hadn't thought of how the masculine presumption of the term farmer is reinforced by legal structures.

"Yeah. But no, I don't write down farmer." She paused to consider. "Maybe in another twenty years I would."

"So you would like, you wouldn't mind, being it." By "it" I meant farmer. Shelly understood.

Judy Beuter Jedlicka, organic farmer and PFI member on her farm near Solon, Iowa, 1999.

"No. I wouldn't mind it. I just don't consider that I do enough farm work to be a farmer."

"Part-time farmer?"

"Part-time farmer, I suppose," Shelly answered. "But once again, Bert's in charge. He's the farmer. I'm the helper. I'm the homemaker. And farm hand. And when that studio gets going, I'll be back to being an artist."

The culture and institutions of farming are such that even a woman who farmed by herself for ten years has some trouble claiming the self of farmer. Clarinda and her husband divorced in the mid-1980s, and she was left with a four-hundred-acre farm with fifty head of dairy cows and a small swine herd, and four children. She kept the farm going for ten years—with the help of her kids, she made a point of telling me.

"We did it all," she explained with matter-of-fact pride one afternoon in her kitchen. "No man. . . . Just a woman and her kids."

Together, they kept the dairy going for about half a year, until her former husband took the cows to his new farm. They kept the hogs for another four years, and the cropland for all ten. During all this time, Clarinda drove a school bus in the morning and waited on tables most evenings.

They had some help from a neighbor, Clarinda also made a point of telling me. "He had a farm on the other side of us," she said, flipping through a photo album as she talked. "And when he was driving by, if he saw me or one of the kids trying to hook up to the disk, and we were having trouble or something, he'd pull in and help us. So he would kind of help us get going or guide us along. But we farmed alone for ten years."

Not only was she farming, she was a "guy," or very nearly one, and dressed that way too. "During those ten years, I had a pliers pouch with a pliers around my waist," she explained. "I just went to town like any of the guys."

She paused, remembering. "I think I probably was the only woman that actually farmed alone without a husband. I think I probably was."

Clarinda didn't feel that she was excluded by "the guys," but she was made to feel her difference. "[During] the about ten years that I was alone farming, I was always invited to the farm meetings. I knew I would be the only woman there, and I went to every one of them, darn near. I'd come walking in, you know, and the guys would kind of grin. And I'd sit right up there like the rest of them. I always found that kind of fun."

"Sounds like you didn't let it worry you at all."

"Didn't bother me at all. In fact, it was kind of fun. And most of the guys—"

"Nobody hassled you or anything?" I interrupted.

"No, no. Not around here. There was only one guy, and he was a good friend of ours. He happens to be a neighbor around here, and we are still good friends. But he was the only one that made me a little bit upset. Because he thought for sure that I would never keep farming after I got divorced. He thought for sure that I would give up my ground. And he made the comment to other farm guys that, 'Well, she would be crazy if she thought she could go out there and farm like that.' Cause it was bothering his manhood as a farmer, I guess."

Clarinda was thus well aware that she was treading a social boundary, and that some might resent it. But she found a way to laugh it off.

"It always gets back to you in a small town, and I used to laugh." She switched to a mock hush. "Behind it all was that he was hoping to be able to rent this farm."

"I see," I said, semi-whispering too.

"I had that figured out a long time ago," she said, laughing. "So I thought that was amusing."

If Clarinda was right in her suspicions, amusing is the polite way to put it. For not only was she made to feel a difference, it was used strategically against her. But Clarinda is polite, and those skills no doubt served her well during those years. She is also remarkably confident in herself, at least looking back at it now.

"I feel like I was born in a tractor seat. . . . I mean, I can run a tractor probably about as good as a man. And I know how to take care of most any livestock, because I did it when I was home with my parents and with my first husband. I just grew up knowing how to do these things. So when it turned out that it was just this woman and kids on that farm over there, I just tried to keep everything going. . . . I guess I was stubborn, and I was determined to keep ahold of the farm."

Clarinda has recently remarried, to Mack, another farmer. They both sold their farms and bought a new one, so as to be on "neutral territory," as she put it. Clarinda still does a lot of work out in the fields, although not as much as she used to. And she says she and Mack share equally in the decision making on the farm.

"It's both of us," she explained when I asked about it. "In fact, if he happens to be gone on the road, and maybe something needs to be done, he has a phone in his truck and we can communicate."

In addition to farming, Mack is a teamster and drives a truck cross-country, as many Iowa farmers do.

"And he may call back and say, 'Well, what's the weather like? What do you think we ought to be doing?' And I'll say, 'Well, probably the north forty needs some chemical put on. And then maybe tomorrow, if you're home, or if you're not, I'll go out and I'll disk some in.' So it's kind of like that. So it's actually both of us. When he met me, he saw that I was fairly self-sufficient in the farming department. So he respects me from day one in that area."

But with all these skills and experiences and her joint role on the farm with her new husband, Clarinda still has some difficulty in calling herself a farmer. It came up when we got to talking about the physical requirements of farm work.

"Farming is hard," Clarinda was saying. ". . . I mean, there's a lot of things that a woman is just not cut out for. Unless—"

"In what respect? Just mean strength?" I wanted to push her on this one. "It's all machines, right? Push the buttons, pull the crank."

"No, it isn't like that. It wasn't like that when I was doing it at all. It's hard to explain."

"Isn't that the reason to have machines? So that people don't do physical work?"

"But machines do not just make everything—" She paused to consider. "It's not like you go about and you're hitting computer buttons, and everything is just happening. And the average woman is not strong enough, strength-wise, to handle a lot of things."

We were, in short, arguing—politely, but arguing nonetheless. In the back of my mind were the anthropological studies I had read about how the average woman around the world does quite a bit more physical labor than the average man does, hauling wood and water, washing clothes, grinding grain, sowing and weeding and harvesting, and more.[9] I agreed with Clarinda that sometimes there are situations where the typical man's greater upper-body strength is a distinct advantage. Yet it seems to me that that is very often because people have developed technology with that assumption in mind. And there are usually other ways around such situations, most notably getting help—as Clarinda explained she often did, drawing especially on that one helpful neighbor.

But Clarinda didn't see it my way. And she seemed to feel that going for help in these brawny moments made her less entitled to call herself a farmer.

"It's hard to explain. Very hard to explain," she said, perhaps with a measure of the exasperation of disagreement.

"Do you consider yourself a farmer now?" I asked, trying to move the conversation on a bit.

"Yeah. I think—" She hesitated, considering.

"Like if you were going to fill out a form that asks what your occupation is?"

"That's what I put down," Clarinda replied, possibly putting her answer in the past tense.

"You'd put down farmer?" I asked, trying to be neutral about tense.

"Yes, I do," she said, certainly in the present tense now. But then she definitely switched to the past tense. "I really put that down those ten years I was alone. No, I guess I say I'm the wife of a farmer. I don't know. Yeah."

"You consider yourself a wife of a farmer now, not a farmer?" I was really pushing her now, I hoped not unkindly.

"Yeah. If you want to get technical, if I was filling out papers, yeah, I suppose."

Those forms again. I very much doubt that Mack has ever considered writing in on them "husband of a farmer."

Rob eased the John Deere into gear, and off we went with barely a lurch. It takes some skill to handle the clutch on a tractor, as the engine speed is controlled by a throttle, not an accelerator, which makes it easier to maintain an even speed once you get going. Rob was clearly good at what he was doing. Given my precarious perch on the arm of the seat, I was both impressed and grateful. But I had a harder time getting in sync with Rob's aesthetic sensibilities.

"Too bad you're out here today when I'm just doing these shitty little fields," he said as the tractor got to speed. "Out on a nice big flat field you'd be able to really see what this thing can do."

The "shitty little" seven-acre field we first did, and the second one not much larger, seemed to me the perfect place for Rob to show off both his own skill and the subtler strengths of the great machine. Yet where I wanted to see the 9300's and the driver's capabilities in a tight corner, his tokens of excellence seemed to be sheer number of acres and the magnificence of massive power under one man's control. It was a boys-and-their-toys thing for us both, but in different ways.

"This thing plants about an acre a minute," Rob observed matter-of-factly, pointing at the dial indicating acres planted per hour.

I nodded. Despite his smooth handling of the tractor, I was feeling pretty awkward perched on the arm of the driver's seat, and I was having a little trouble getting the conversation going. I was also still trying to get my bearings, as I haven't spent much time around tractors.

He seemed to sense the gap. "I'll tell you, I don't think there's anything much different about farming today," he offered, trying to help me out.

"Really? I'm surprised to hear a farmer like you say that."

"There's nothing different in what I do." He mentioned a book on agricultural history that I'd never heard of. "It shows how a hundred years ago there were fifty-thousand-acre farms."

"But that was probably in dry areas." I was feeling argumentative, which is not a recommended approach to interviewing. Rob didn't seem to mind, though.

"Yeah, it was probably wheat. But there have always been big farms. Farming's always been the same. There's progressive farmers and farmers who aren't."

"Well, what do you mean by a progressive farmer?" I asked, still feeling argumentative.

"You know, someone who modernizes and invests in the latest technology and improves their farm."

That one shut me up again. I couldn't think of where to go from there without being too threatening or challenging. It didn't help matters that I misspoke when I finally thought of a question. By accident, I started by calling him Vern, the name of the employee whom he had cussed out over the thread sealant.

"Don't ever call me Vern! Call me anything but that!" Rob was plainly hot over that one. I dropped the question. We lapsed back into silence. It was very close quarters for such an awkward conversation. Moreover, I was not the most sympathetic of observers with regard to Rob, and I am fairly sure he recognized that.

"You know, I'm really impressed with how you can tell just where you've been," I said eventually. This was true. I was genuinely struck by how well he could use the outrigger devices on either side of the planter to keep his place in the field. To my eye, it wasn't easy to see the difference between rows that had been planted and rows that hadn't. This is an expected skill of any experienced grain farmer, though, and I think I sounded naïve as much as complimentary.

"Well, I don't want to say that I don't miss a few places," he replied, after a sidelong glance at me.

We came up to a grassed waterway, a common erosion control structure on Iowa farms. I expected Rob to do one of his on-the-dime turns right at its edge, but Rob plunged the tractor forward. The wheels started to slip in the mud and wet grass. The dial indicating acres planted per hour started to drop.

"Come on, baby! Come on!" he shouted. So many acres, so little time.

It didn't look to me as if he was doing the grassed waterway any good either, but I didn't say anything. He was planting along the contours for the most part, however.

We got to talking more when we got to the bigger field. "So Rob," I asked, "did your family help you get going, to get this big operation going?"

"That's the thing I'm most proud of," he said. "No. Everything you see out here, I worked for. I got it myself." He paused to lift the outrigger as he came to the edge of the field. "I was very driven when I got going. I hated to sleep."

"You hated to sleep?"

"That's right. I like to sleep now! But then I hated it. I just couldn't understand why I had to sleep. It was lost time when I could have been out there working. I'd sleep maybe two hours a night, or just lay down in the barn for a hour or so. And then be back up again. I couldn't do it now. But I'm glad I did it then, 'cause that's what got me going to where I am now."

I was feeling a bit seasick, bouncing up and down on the arm of the chair, and was having trouble focusing on the conversation. Rob kept talking without much prompting from me, though.

"Nobody would give me a loan. I got in a car and I drove everywhere, looking for someone who would loan me money. I drove all over the countryside for weeks, sleeping in the car. Finally, FHA gave me a loan." He meant the old Farmers Home Administration that for years provided loans to higher-risk farmers, until it was folded into the new Farm Service Administration of the USDA in 1994, which still provides this service. "They were horrible to deal with and I'm glad I'm done with them now. Don't get me wrong. I'm real glad they were there. I never would have gotten anywhere without them. They got me going. And at the end of that year I had three thousand dollars left over, and I felt real proud. Seems like nothing to me now."

Maybe he could accept the reality of help, at least to some extent, I thought. "So, how did the farm crisis affect you?" I asked.

"Yeah, that's another thing. All this was happening during the farm crisis. I couldn't understand why all these farmers who were in much better financial shape than me weren't snapping up the land. Land had dropped to about seven hundred an acre in the mid-eighties, but they weren't buying. It seemed stupid to me. I bought a three-hundred-acre farm in 1984, and by 1986 I had about twelve hundred acres. That was my first real good year. I had a hundred thousand dollars left over at the end. I got away from the FHA and switched to local bankers. And now I've left them."

He was thoughtful for a minute. Neither of us spoke. He had to swing the tractor sharply at the corner of the field and concentrated on that.

"But you know," he said. "I don't like talking about this stuff. I really don't."

To put it simply, there are a lot of guys in farming, and there's a lot of guy stuff too. But the dynamics of gender were not all that was going on during my visits with Rob, Dale, and Shelly and Bert, and during Sue's visit with Marlan and Kelli. These are men and women, and the gendered aspects of their selves are intimately bound up in their experiences of farming. But these experiences have other, often interwoven, dimensions as well.

Take Rob, for instance. I suspect that much of what being a six-thousand-acre farmer represents for him is class mobility, not masculinity alone. He has a high school degree, but no college. His wife Mindy has a one-year degree in cosmetology, but nothing beyond that. Yet when Donna asked them about which social class they would place themselves in during her interview with them, here's where Rob placed their family.

"Our living standards," he considered. "I'm sure we're in [the] upper class, the way we live. If I'm reading your question properly. We're able to live the lifestyle. As far as how we live, it's just whatever we want. We never have to worry about what we want or what we want to do. I assume that's something that not everybody can do. I figure that would put us in the upper class."

Mindy placed them in the "middle." Still, she feels she's risen beyond where she once was—or perhaps it's more accurate to say that Rob wants her to think so. Earlier in the interview, Donna had asked Mindy about her plans for her cosmetology training. "Do you see yourself going back to doing anything like that after the kids are older?"

"Well, I had always wanted to put one in my home," Mindy explained, meaning a beauty salon. "I always thought that would be ideal. I put my license . . . you can put them aside for ten years and you don't have to go to classes and pay your renewal every year for your license. It's been just about—"

"Mindy," Rob interrupted. He didn't like where he thought this was going. "Do you really think you want to do that here down the road? No. She'll never cut hair again."

"Rob!" Mindy said sharply. "I'm not done."

"Okay," said Rob, chuckling.

Then Donna laughed a little bit, too, trying to smooth things over.

"It's been just about that long," she continued as before, "and I'm thinking now, you know, I'm not near as interested in it as I used to be. 'Cause I was really wanting to put one in my home. And, now I'm pretty sure I probably never will."

This little victory for Rob (if that's how he saw it) was evidently important to his sense of manly success and control within the family. His wife will never have to cut hair again. But he evaluates his masculine achievements in part with the yardstick of class attainment, through putting cosmetology behind them and being able to do "just whatever we want."

Class, in this strictly economic sense, is but one metric by which farmers judge each other and themselves—not only as men, or as women, but as worthy members of rural society and society at large. It is but one metric by which they judge something broader: status.[10] What these metrics are was not always the easiest thing to get at in interviews, though. "So, how do you folks measure status out here?"—that's just not the kind of question you ask people who have invited you into their home. Status is usually contested, often subtle, and normally unspoken. Of course, rural people do sometimes speak of it, but among the subtleties of social life are knowing when and how to speak of it, lest one's contestation of status backfire. You can't just claim status or summarily deride someone else's without risking your own. Status is something others confer. And those others locate themselves within their own relations of support and contestation for the claims of status. So it is easy to offend. You have to be careful in rural Iowa, just like everywhere else, not to snarl up the web of honor as you step along its sticky threadwork.

Like everywhere else: My point here is not that farmers are all Machiavellian, or at least not unusually Machiavellian. As with power, with which it is closely connected, some degree of status is central to social psychological health, of course. We should not lament that people desire it in rural Iowa, too—although, at least in my view, we should indeed lament that the people of rural Iowa do not always seek a democratic distribution of status. Like everywhere else.

One common way that ideas of status emerged was in discussions about the neighbors, the ever-watchful neighbors. "I'm living in a fish bowl," was how Bert put it. And a lot of farmers feel that way.

Luverne is one. He's in his late fifties and has been farming since he left high school. He's up to almost seven hundred acres now in corn and beans, with some livestock, some of it owned and some of it rented. It's a family operation through and through. He farms with his son, rents some land from his father, and the heart of the operation is a farm he rents from his aunt. That's where he and his wife live, on a small parcel that his aunt sold them outright, and have lived since 1968.

"So when you came in, in '68," I asked him, "had you been doing anything else before that?"

"Yeah, again, I was farming with my dad down toward Powell Center."

"Oh, yeah." I'd forgotten that he'd mentioned that a few minutes earlier.

"We farmed for him—we were married in 1963," he continued, trying to remember. "And I farmed down there with him until 1968, when this farm become available. I kinda had my eye on the farm up here because it's been in the family a long time. I was young. I needed to go someplace because I had another brother behind me down there. So it was only natural that I would come up to this particular farm. The people who were here prior to me, they lived here for probably thirty years. And it was quite a decision for her to let those people go. She gave them two years' notice. He found a farm. He moved off and I moved on."

"Uh-huh," I nodded.

"I was young, and I know the neighbors were really watching at that time—the older neighbors. They probably thought, 'Well, here's some kid coming in here, and things are gonna go to the devil.'"

He smiled, and I gave a little laugh.

"But we gradually worked at it. And we were successful the first few years. In fact, I had some of the older neighbors stop in and say, 'The crops sure look good.' I think maybe I surprised them. But I knew people were watching me.

So I had to do a good job. Because—" He paused, and took a bit of a deep breath.

"I was kind of on the spot in a way, you might say," he said finally. "People driving up and down the road thinking 'Well—' Because the past tenant was here for so long, I could see the neighbors talking. 'Here comes a kid in here, and he doesn't know nothing.' "

There is a lot of pride behind Luverne's words as he discusses how he thinks maybe he surprised the neighbors with how well he did. He doesn't know for sure. (The status you have in the eyes of others is something you never know for sure.) But he thinks he contested their expectations of a young kid, and he succeeded. Things didn't go to the devil. The neighbors even said that the crops looked good, lending external credibility to his claim for status. This is a masculine pride—note his frequent use of the singular "I" and "he" to describe his accomplishments. And, although he doesn't say so directly, it can be inferred from Luverne's remarks that he was making money. I imagine he was well aware of that potential inference.

But Luverne wasn't just trying to demonstrate his masculinity and class with this story. He was also trying to say that he's a "good farmer"—a common phrase that runs through rural talk in Iowa. And he was also talking about the visibility of all these statuses to the neighbors driving up and down the road.

We asked a number of farmers about how they judge what a good farmer is, and they consistently expressed their answers in visual terms. We've heard from Rob a lot, but let's hear from him one more time.

"I'd have to honestly say that I'm pretty much the oddball around," he replied with a laugh to Donna. "I mean, a good farmer—most all my neighbors I really feel are better farmers than me."

"Really?" said Donna, with genuine surprise. She wasn't expecting humility from Rob.

"Yes," he continued. "I'm in a area where they are really good farmers. The crops always look real nice. I've always got to screw up somewhere, or something ain't right. I'm still waiting to where I can get everything right on all my acres. On the general, it's okay. But I always, it seems like, got a field or two where I do something wrong. What I'm really striving for is to have everything one hundred percent, and that's hard to do. It is for me. I can't do it. I look at everybody you see that does a good job of farming—"

"What, is that when you drive by their farm?"

"You can tell by looking at a crop what a person is doing."

"What are you looking for when you look at that crop?"

"You can drive by a field and look at it, and you can tell by looking at a field of corn. You can look at the stand they got, and the evenness of it, the color of the corn, the way it's planted, the weed control in the field. All them things. When everything is right, that's the most pretty thing."

Donna nodded. She wasn't all that surprised by what Rob was saying. Farmers have a reputation for the aesthetics of the weed-free field, like some vast suburban lawn: Chemlawn agriculture. They are increasingly criticized for it too, and there has been something of a campaign by extension agents and by sustainable agriculture advocates to get farmers to think about the money they're wasting in getting those last few weeds out, not to mention the environment they're wasting. Donna was curious to know what Rob thought about how far one should push it.

"What about weed control?" she asked. "How tolerant should a good farmer be on weeds?"

"That's kind of a loaded question." Rob was clearly familiar with the issue. "You think a good farmer has spotless fields. Which is what I like too. Sometimes it just don't happen, though. What I am trying to say is, there can be the good farmer who wants a picture-perfect crop. But there can be a good farmer who's looking at the bottom line—who's gonna grow just as good a crop, but maybe have a few weeds out there. Both of them may be good farmers."

Donna nodded.

"One may be looking at the bottom line a little more. It's a gray area. . . . That's my own personal view of it, I guess. I mean, I want things picture perfect. We're getting into some grass problems, and they're developing new chemicals now, but this cup grass has been really hard to control." This is the same weed problem that Rob had mentioned to the passing fisherman.

"A lot of people that are really good farmers, and always have things right," he went on, "even have that coming through once in a while. It gives you an idea of how hard it is to control."

Although he recognizes that the bottom line does put some constraints on how far you should go with it, for Rob a picture-perfect crop still means a picture-perfect farmer. Not only does he imagine that others are inspecting his fields, he is constantly watching theirs when he drives by. Which is why he imagines other farmers watch his as they drive by. And they probably do.

But it is not just a matter of the suburban lawn ethic writ large, of rural life imitating suburban artifice, just as suburban lawns themselves imitate a pastoral understanding of the farm.[11] There is the same visibility of the self at work in suburb and farm, but the degree of what is visible, and thus what is at stake, is so much greater for farmers. For all its Jeffersonian individualism and privacy, farming is a uniquely public occupation.[12] Your neighbors, your friends, your competitors, and anybody else who understands farming can see so much of what you're doing. "I was kind of on the spot," said Luverne. The farmer is constantly on the spot, that public spot of self, the farm. I didn't have to constantly hang my emerging draft of this book on my office door, or on the front door of my house, and I'm glad I didn't have to. But, in effect, farmers do. It really is a fishbowl.

There is also an ethnic dimension to farming in Iowa. I don't just mean the continuance of ethnic enclaves in the rural landscape, like Elk Horn, Iowa, with its Danish National Museum, or the continued celebration of Dutch and German ancestry in many of Iowa's small towns. Most Iowa farmers are indeed well aware of their ethnic background and the background of other farmers and rural people. (We asked about ethnicity with almost everyone we interviewed; few had trouble answering the question.) Almost all Iowa farmers are white, and they recognized that too, when it came up. On several occasions, we heard discriminatory comments about blacks, Jews, and Mexican migrants and immigrants; I have reported on one instance of that already. Yet, on the whole, ethnicity seemed mainly a matter of identification, not social mobilization, as far as farming is concerned.[13]

But that's not all I mean about an ethnic dimension to farming. We were also impressed with the way that farmers and others associated with rural life in Iowa often described farming itself almost as a kind of ethnicity—as something that one is born into, that one should marry within, that one hopes one's children will perpetuate. In these ways farming becomes more than an occupation, and more than a culture associated with an occupation: It becomes part of farmers' bodies, the history of their bodies, and the future of their bodies. And it becomes as well something that is connected with a community of other bodies with similar histories and futures.

Ethnicity may not be the best word for it. Among other reasons, there is considerable disagreement about what ethnicity is. Ethnicity concerns ideas about cultural differences, and race concerns ideas about physical differences, suggested the sociologist Pierre van den Berghe, in an influential formulation.[14] More recently, others have suggested that both ethnicity and race are about culture, for the physical differences that some see in race don't hold up when one looks closely at the matter. (For example, why is the child of a "white" parent and a "black" parent typically regarded as "black"?) But, on the other hand, most groups imagined to be "ethnic" are also imagined to be distinct physically—perhaps by differences in hair, skin, and facial structure. And on top of that, races and ethnic groups come and go. (For example, who speaks of the Irish "race" today, as was common in the nineteenth century?) In short, it's a conceptual mess.[15]

But running through most people's conceptions of race and ethnicity is a common form of social imagination: the cultural identification of an imagined community of descent or heritage—what I and my colleague Iverson Griffin have elsewhere termed *heritas*.[16] The emphasis here is on imagination. Heritas does not depend on objective reality. Yet the heritage groups that form from heritas are no less socially real in their consequences for human lives.

And such a process of imagining and identifying a community of descent seems very much at the heart of farming culture, too. Take the following familiar phrases: Born on a farm, grew up on a farm, family farms, farm families, home farm, country boy, country girl, I'm a farm girl born and bred, I'm a farm boy born and bred, this farm has been in the family since. . . . Or, as Mindy said to Donna about Rob's passion for farming, "It's just like it's in his bloodstream." The daughter of another farmer explained her whole family's passion for farming in virtually the same terms. "Oh, I guess it's just in our blood."

Note the persistent familiarity of these familiar phrases—that is, their persistent reference to matters of the blood, of descent, of family. Just metaphors, perhaps. But even if rural people intend them as only metaphors, they certainly are well aware of their cultural power. And that's a form of imagination and identification in itself.

It's the kind of thing that is most evident at its social boundaries. At least Richard seems to find it that way. I got to know him at a field day I went to at the farm of Dave Lubben, a sustainable farmer well known in Iowa and a former president of Practical Farmers of Iowa (about which much more will be said in chapters to come). Richard has a small farm in the southern part of Iowa, where he is trying a number of sustainable farming techniques. Richard was particularly interested in rotational grazing, something Dave Lubben has had going for some years. Richard had traveled a couple of hours to get to the field day.

I wound up sitting next to Richard on one of the hay racks Dave was towing behind his tractor to transport the tour group around the farm. It was a beautiful day, and Richard and I chatted some as we bumped from stop to stop. Actually, Richard was a lot more interested in talking to Dwayne, another sustainable farmer, who was sitting on Richard's other side. I was sort of part of their conversation, and sort of not.

"It's a pleasure to be out here," Richard said. "I've got no one to talk to in my neighborhood. They all think I'm crazy—my problems are my problems. None of them are doing what I'm doing." This sense of aloneness is something that sustainable farmers often feel.

Dwayne nodded in knowing agreement. "You can bet they're watching, though."

"Oh yeah," replied Richard. "I just tell them I don't know what I'm doing! I wasn't born on a farm, you know. I'm a city boy. So I just tell them I don't know any better!"

Richard laughed hard at his little joke, but Dwayne just smiled, and repeated, "Yeah, they're watching what you do all the time."

"And saying, 'Yup, he'll be gone in a year or two.'"

They both laughed hard at this one—the gallows humor of those who are "still in," as Iowa farmers say. There was a pause in the conversation, and I asked, "So is that really true about the neighbors, and the pressure?"

"Oh yes," said Dwayne seriously. "That's your brother-in-law, or your friend, across the way. It matters to you. Yeah, it's hard."

They went back to discussing rotational grazing techniques. Apparently Dwayne and Richard didn't know each other very well. But I think Richard didn't know anyone else there at all. He seemed a bit over-eager in his conversation with Dwayne. Although he was a farmer, I think he thought of himself as an outsider to the group at the field day, probably in part because he came from a different region of Iowa. Whatever the reason, he seemed to feel himself at a social boundary and couldn't let the "city boy" issue go.

"I've got a lot to learn about driving cattle," Richard said to Dwayne later on. "I went to a talk by"—he mentioned a famous grazer, but I couldn't remember whom when I wrote up my field notes later—"and learned a lot about how to approach the cows. About their blind side, and getting them to come toward you and go where you want. Being a city boy, I didn't know any of this stuff. You probably always knew all about it."

By pushing the city boy/farm boy boundary a second time, I think Richard was, perhaps not entirely consciously, asking for a clear sign of acceptance. Dwayne gave it to him, saying, "Oh, I don't know. You've probably got an advantage, because you don't have preformed views."

I was probably almost as glad to hear Dwayne's response as Richard was. I often felt myself the "city boy" during the course of the field work. That same afternoon, as we were riding on the hayrack, I pointed out a field we were passing and mentioned to Dwayne, trying to sound knowledgeable, "Boy, he's sure got a lot of button weed in that field."

Dwayne glanced at me and then drawled, "Ah, around here we call them soybeans."

We had a good laugh over that one—soybeans are one of Iowa's two major crops, after all—and I got a similar response when I told the story on a couple of occasions to other farmers. I found it sometimes helped to make a joke of my outsider status, raising and partially dissolving it at the same time.[17] At least it made me feel more comfortable.

It's not that I was ever sneered at for my urban background, at least not to my knowledge. But it seemed a question that was always being asked. Farmers would usually find some polite way to get at the issue, often in those so-where-are-you-from questions that are common when two people first meet. Usually the issue was got at indirectly, passively, but sometimes it came straight on.

"What's your home town?" I was asked by a farmer at another field day—not where, but what. Locals know the where.

"You mean here in Iowa?" I answered, feeling myself at the borders already. "Well, actually I come from New England. I was born in Providence, the capital of Rhode Island."

"So you're a city boy?" came the immediate response.

Perhaps I was being hypersensitive, but that unnerved me, as did several similar exchanges on other occasions with other farmers. True, I already felt marginal as a sociologist studying agriculture. There was nothing necessarily about descent in any of these comments. Possibly they really were just interested in where I was raised.

But I'm pretty sure there was more to it—more to the Iowa custom of awarding a "century farm" plaque to any farm that's been in the same family for at least a hundred years, more to the terrible sense of loss many farmers feel when their children all marry or move out of farming, more to the common sense of bonding that farmers express when they meet another farmer from another region for the first time.

And I'm pretty sure that Emmet thinks there's more to it, too. Emmet and Brenda have a four-hundred-acre farm in Powell County, where they raise grain and also a small sheep herd. They live in a beautiful old Victorian farm house with a sweeping front porch and a couple of glorious oaks in front. He does more of the field work than she does, but Brenda is heavily involved in all aspects of the farm operation, inside and out, although "she's from town," as Emmet explained as he showed me around the farm.

Emmet's not from town. They live on the very farm where he was raised, in fact. It's not a century farm, but he's the third generation of the family to farm it. And he loves the old place. He seemed to take particular delight in showing me the barn. It didn't look like much to me until we got inside, as the outside is now metal-sided. (I even thought it a modern building when I first saw it.) But inside the old mortise-and-tenon beams were all in glorious evidence. The barn is still in active use. They keep their tractor there, as well as a sizable store of square bales for the sheep.

They had just finished a cutting of hay and he needed to stack some more of the bales. I gave him a hand with the pulley, and we talked. Their income isn't high, apparently, and they're pretty worried about it. They own only a third of the farm themselves. The other two-thirds are owned by Emmet's two siblings, and Emmet and Brenda rent from them at a preferential rate. The siblings love the farm too, and they cut their brother and sister-in-law some fiscal slack. Nevertheless, Emmet and Brenda are running close to the edge. Emmet made some jokes about it, but he was plainly worried.

We finished the last bale. Emmet wiped some sweat from his face. He looked at some of the old timbers and said, half to himself, "I can't lose this place."

Then he looked at me. "This is one of the things about a third-generation farm. If I lost this place, my name would be mud."

But at some point we have to call a halt to all this sociology and recognize something obvious—something so obvious that it took me a long while to recognize that I recognized it: that farming is a buzz, a phenomenal buzz. Let's face it, given the puny returns on the amount of capital invested, it can hardly be said that very many Iowa farmers do it for the money. At least, it isn't exactly the kind of enterprise that business school training would recommend to the avaricious. Which, perhaps paradoxically, is one of the main reasons why family farms still exist at all: What big corporation wants to step in where the margin is so small and so uncertain?

No, farmers do it for the buzz. Even the Big-Time Operators do it for the buzz. (They too could be making a better return in some other enterprise.) They do it because farming is an all-consuming endeavor, a great wave that catches a rider and carries him or her along in its frothy excitement, thrilling in one moment, threatening to tip the board over in the next, and nearly always exhilaratingly exhausting. Sometimes it's just plain exhausting. Sometimes it's deeply discouraging. Sometimes the waves are too unruly and the board breaks. Sometimes you feel like you can't get out of the surf. But what a rush when you catch it right, and what a rush it is to try. So many acres, so little time.

Or perhaps it would be better to say that this buzz is where the sociology really starts, where it comes to life. For this is a phenomenological buzz too. It's a buzz that comes from the intensity of the sociological relations that pervade every second of farming life: gender, class, status, ethnicity—all big buzz words of contemporary social life. It's the buzz of self. Farmers typically don't experience what they do in these terms. They experience these relations tacitly, in the main, as they contend with broken planters, stack hay, relish an indoor hog medicator, check the latest commodity prices on the computer, admire a fine herd of cattle, and race to get the crop in ahead of the rain. It is, as I say, a phenomenological buzz.

I can't give a single quotation from our interviews to prove this, however. The importance of the intense *phenomenology of farming* is there in nearly every quotation, at least to some degree. That's the way phenomenology is, unseen but everywhere. Let me put it as simply as I can: These people are really wrapped up in what they're doing. It's farming's vivid capacity for bringing together so many strands of social life that makes it that way. So many strands, so many acres, so little time.

Seen from this perspective, it may often—or even usually—not matter that the structure of agriculture in Iowa undermines economic, social, and environmental sustainability. At least it may often not matter for those farmers who are "still in." Sometimes it does matter to them, and very much so, as we have heard. But the struggle to create a stable identity against the backdrop of uncertainty becomes an ongoing accomplishment that provides meaning to the lives of Iowa farmers. The roiling structure of agriculture thus becomes, perhaps paradoxically, a means to meaning. Especially in conditions of uncertainty, the accomplishment of meaning is a hard thing to give up. And so farmers farm on, if they can, helping reproduce the structure of agriculture, with all its uncertainties, as they strive for a perch on the noisy tractor seat of personhood. For, to repeat, it is not corn and beans that they are really trying to grow, but their own selves.

The metaphors and figures of speech have been flying fast and furious, I know. But suffer me one more go. As silent as the springs may increasingly be in rural Iowa, the buzz of self is as yet plain to hear.

5. Farming Knowledge

Gaylan is sixty-one. He's been farming since he was sixteen, and he's still hanging in there. But just barely.[1] His daughter and his son-in-law have already had enough. Two years ago, they were all farming together in Powell County, a thousand acres, in fact. Now it's down to Gaylan, and probably not for much longer.

Or rather, two years ago they were all part of the same farming operation in Powell County. Very definitely, they were not farming together, as Gaylan's daughter, Renea, and her husband, Kurt, explained to me one spring afternoon. We were sitting in the paneled dining room of their Victorian farm house, as they retold the story of the hats-flying emotional roller coaster of the last few years. How they came back to Renea's family's farm when it was about to collapse. How they struggled to bring some new ideas into the operation. How they both left their urban careers behind. How excited they had been to be finally living their dream: farming. How it had all come apart. Their words tumbled out, one on top of another.

"We got back here and we didn't—" Kurt had begun, when Renea threw him a mock elbow. "What?"

"Here, let me tell the story," Renea interjected.

"I'll tell it!" came back Kurt immediately, even before Renea had finished her interjection. "I'm more succinct," he teased, a big smile in his big beard. Renea shot him a glare. Kurt relented. "You tell the story," he went on, waving his hand.

"You, you leave out details," Renea teased back.

"I don't either!" Kurt replied, clapping his hand to his chest and pretending to be deeply hurt by the suggestion. Renea and I roared with laughter.

But for all their good humor it was serious stuff, as they both described it, the baton of the story line often passing between them, albeit sometimes with a bit of a yank. It's a long story, but it's worth hearing.

Although they had lots to say about themselves, Gaylan was always present in the tale, the center pivot of the plot—the living ghost of their own King Lear. After all, along with Renea's mother, Gaylan owns the four-hundred-acre core of the thousand-acre farm, and rents the other six hundred, much of it from family relatives. And despite the co-ownership, he's the "farmer," as Iowans usually understand the matter.

"I think Dad's dream would have been to have all of his kids farming with him," Renea considered. By "his kids" she meant herself and her husband, but also her sister and her brother-in-law, who farm on their own in a different county. "I honestly believe that."

"Well, he said that," agreed Kurt.

"Yeah."

"And I think he was being honest. He said that a couple of times."

But it's hard to farm with your kids—or anyone else, for that matter—unless your understandings of what farming is, and who farmers are, coincide, or at least complement each other. Because not only do farmers farm the self, they farm knowledge.

"The paradigms that we're operating in and the paradigms Dad's operating in are just—" Renea searched for the word.

"Incompatible anymore," finished Kurt.

"So far apart," Renea added.

Renea and Kurt are big advocates of "holistic resource management," an approach to farming that is increasingly popular among those who identify with sustainable agriculture. Almost all its adherents refer to it mainly by its acronym, HRM, or more recently simply HM, for "holistic management." HM emphasizes the importance of three main factors in decision making: land, money, and people. In this it closely matches what is frequently described as the "three-legged stool" of sustainable agriculture: environmental, economic, and social sustainability. But HM is distinctive in giving people concrete procedures for working all this through in their lives and on their farms. Basically, HM tells people the following: To be excellent record keepers. To think about what your goals and values are. To figure out, through those records, whether what you're doing matches your goals and values. To discuss all this constantly with everyone in your family, men and women, adults and children alike. And, most important, HM tells people to think holistically, as its name suggests—to see interconnections and contradictions, and to think about a family and farm as a system. Professional HM trainers give day-long workshops, even week-long workshops, across the Midwest and increasingly elsewhere too. Thousands, by now, have taken them.

HM was first developed and popularized among farmers by Alan Savory, a wildlife biologist and rancher from Zimbabwe, and there is now an Alan Savory Center for Holistic Management. "Just by weird circumstances, I ended up going to one of Alan Savory's first courses he did," Kurt recounted. "I was just so excited because it made sense. It tied things together. There were pieces missing, and this started tying all this stuff together. . . . I'm a true believer," he concluded with a laugh.

PFI members Leo Schultes and his sons, Ryan and Tony, bale hay near Audubon, Iowa, 2001.

Both he and Renea have been taking HM courses ever since, part of their long-term plan as a couple to leave their city jobs and return to farming. (Like Renea, Kurt was raised on an Iowa farm.) They had come close to buying their own place in Wisconsin, when Gaylan invited them to join the troubled family operation in Powell County. They agreed, with reservations, but central to their decision was that they had persuaded Gaylan to institute holistic management. In fact, the family had begun HM a couple of years before Renea and Kurt returned.

"Mom and Dad took the [HM] course, and we were driving back and forth," Renea explained.

"We'd come to these team meetings too," Kurt added.

The HM team meetings weren't going all that well, though.

"They would just be like, you wouldn't get anywhere," said Renea.

". . . They would sit down, and they were going to discuss something. And then they would have a two-hour break to try and figure out how much money they spent on soybean meal in the last two months." Kurt whistled through his teeth, remembering his frustration.

"Because nothing was on computer," continued Renea. "There were no records and all that kind of stuff."

"I see. Shoebox record keeping," I said.

Kurt nodded.

"All this economic stuff was supposed to have been done prior to the meeting," Renea went on, "but Dad never got around to it. It ended up just being very frustrating fiascos. It was very obvious that Dad was kind of confused and feeling rather powerless, I think, by the whole situation."

Despite Gaylan's evident lack of enthusiasm for HM, Renea and Kurt nevertheless took up her dad's invitation to come back to the farm. They were excited about what they saw as the family progress, albeit limited, that had already been achieved through HM. They hoped that, in time, it would lead the farm away from the standard Big-Chemical and Big-Tractor approach that Gaylan had been using for years.

Their hopes were further fueled when Gaylan agreed to cut back on herbicide use, with hopes of saving thousands of dollars a year. Renea and Kurt thought they'd finally had a breakthrough with HM. But it turned out to be a disaster. It is indeed possible to cut herbicide use well below what most Iowa farmers deem necessary, but not the way Gaylan did it.

"Instead of going out and talking to the people who really know how to do it—and I mean, I've studied [HM] for ten years," Kurt recalled. "I did research with farmers. And you have to have a system. You can't just say I'm going to cut herbicides. You've got to have a whole—"

"Right, right," I said. I meant it. I'd been to Dick Thompson's farm. "Right, right, right. It's a larger change."

"Right. Gaylan didn't do that. He just started cutting herbicides."

"Cutting rates," said Renea, meaning application rates.

"Well, guess what?" Kurt went on.

"Weeds!" they both exclaimed together.

"And Dad is obsessed by weeds."

One of the ways that sustainable farmers cut their herbicide rates is by using a "bander." Instead of carpeting the entire field with chemicals, a bander allows farmers to apply herbicide directly to the rows, where the crop is, while taking out the weeds in between the rows with a mechanical cultivator. There are also ways to take care of the weeds in the rows without chemicals; the ridge tilling used by Dick Thompson and others is one technique. In ridge tillage, you grow the crop on low ridges, as the name implies. During planting, a special planter shaves off the tops of the ridges and tosses the removed soil into the little valleys in between, carrying with the soil any weed seeds that were lying on the ridge, and covering up any weeds that have begun to grow between the ridges. Here in this tumble of loose soil, the weeds have trouble germinating. Some will, nonetheless, and the farmer later makes a pass or two with a cultivator to throw the loose soil and any sprouted weeds between the rows back onto the ridges, rebuilding the ridges for next year and also covering up any weeds that have begun to grow on the ridge tops. (By this time the crop has sprouted high enough that it doesn't get covered over too.) It's a bit complex to visualize, but the result is that the row area (that's the top of the ridge) and the area between the rows (the little valleys) each get at least two treatments for weeds: an uprooting and a covering up of what's there. Usually, it means there is no need for any herbicide at all.

Still, banding by itself allows for a big cut in herbicide use. (But you'd better have a plan for how you're going to handle the weeds where the bander didn't get them.)

So, I asked, "Was he banding?"

Kurt rolled his eyes. "Gaylan's got a brand new bander that's only been used, um, I don't know, on maybe a couple hundred acres. He had all kinds of disasters because he didn't develop a *system,*" he said, emphasizing the word.

And to develop a system, say Renea and Kurt, you have to do all that record keeping HM talks about. Shortly after they arrived, Renea and Kurt put the farm's finances on a Mac computer that Renea had at home for her consulting work in human relations for small businesses. Kurt ran the numbers one night and saw a train wreck just ahead.

"You had figured out that we were going to have this major financial crisis in about the beginning of July," Renea said, looking over at Kurt. "And Dad didn't want to deal with it."

"Gaylan wouldn't look at it. It was like 'No, no, we're not going deal with it.' It was about August 1st. No, it was July 1st."

"It was July 1st," Renea confirmed.

Kurt nodded at her. "'We're going to be in trouble,'" he said, trying to recapture his words to Gaylan. "'The operating note is going to be used up. There's not going to be any income. We got to deal with this. What are we going to do?'"

Then he switched to a deep, growly voice, his imitation of Gaylan, edged in some anger. "'Nope, nope, we just got to farm. Got to get working. We're taking up too much time. We could get some work done if we didn't have to sit around and talk, dammit! Let's get out there and work.'"

There was a train wreck, though, and it probably didn't help that Kurt had predicted it. The banker called them in.

Dean Hansen, PFI member, in his pickup at his farm near Audubon, Iowa, 1999.

"There was no money," Kurt explained. "And the banker said, 'Something's got to give. Somebody's got to leave.' And so we were asked to go."

"That was probably the hardest thing Dad has ever had to do in his life. For a while there I wasn't sure if he was going to make it."

"He was in bad shape. I thought he was gonna—I hung around with him, I remember, for a couple of days, just kind of never left him. I just didn't know if this guy was going to make it. If he'd fall off and just break his neck, or forget he was standing on top of something and fall over. He was just totally out of it. He was so stressed out from it."

"It just tore him apart to the very core."

Kurt poured himself another cup of herbal tea from the pot we'd been enjoying together. "There are exciting things you can do with row crops. I get excited about alternative ways of doing things—that would make money! And also would fit with more of my values towards the land. But there ain't no way. I mean, Gaylan will not consider [it] . . ."

"And if you go to him and say to him," I asked, " 'Look, here's an alternative. I've seen it. It will work.' "

"He doesn't believe you," answered Renea.

". . . So, we weren't able to do anything alternative," Kurt explained, running his hand through his beard. "And I mean alternative to make money. I don't mean alternative just—"

"For being weird," Renea inserted.

"For being weird," Kurt agreed. "I mean, where we could have increased profits. He'll spend thirty grand at least on herbicides this year. Half of that, without any risk, could be eliminated. Then you've got the bander sitting in the shop. We already *own* the bander. You could take the bander and you could cut that in half again, almost. You really could! But it just ain't going to happen. There's no way. He's burned out, and by God he's going to farm the way he knows it works."

Or rather the way, at some level, he knows it *doesn't* work. But it's the way he knows, and thus the way he knows himself. Even when things are going badly, even when the banker jabs a finger at a key clause in your loan papers, it is no easy matter to change what you do. Because it is through knowledge that farmers farm the self. If the self is the real crop of farming, knowledge is its real soil, each producing the other in a constant cycle of social nutrients. And, as any good farmer knows, to get a good crop you have to tend to the soil as much as to what it grows. You have to cultivate them both, together. To cultivate knowledge is immediately to cultivate the self and its social relations, and vice versa. And when I speak in this book of the *cultivation of knowledge* and of *knowledge cultivation,* I mean to refer to them both—to knowledge and to its active culture of identity.[2]

So when Gaylan says he doesn't believe in an alternative that Renea or Kurt suggest, one they have learned from others, he is doing more than disputing facts. He is rejecting those others and the social ties that Renea and Kurt have with them. He is also rejecting the manner of self that these relations cultivate. And something more: He is rejecting Renea and Kurt and their own manner of self.

Which hurts. It hurts Gaylan, for ideas like the importance of careful record keeping, banding herbicides, and lots of discussion with others about how to farm threaten how he has come to understand his self. It hurts Gaylan that his own daughter and son-in-law would confront his self in this way. It hurts Gaylan to have to reject his own daughter and son-in-law. And it hurts Renea and Kurt to be rejected by him, because they find their selves in part through him as well.

I hope our interview gave Renea and Kurt a chance to work through some of these issues. I think it did. In any event, I think Kurt was seeking some emotional release when he told me the following story about Gaylan.

"I got to tell this story," he began. "This is cruel, but I'm telling it anyway. The one neighbor, who we told you to try and interview, he was over and we were talking—"

"Which neighbor is this now?" I asked.

"Because they've talked to a bunch of people already," clarified Renea.

"Oh, you have? Um, um, um," Kurt searched for the name. "Rex. Have you talked to Rex?"

We had. Sue had interviewed him the day before. Rex hadn't told us, but evidently he has been experimenting with some new sprays that are supposed to allow a farmer to cut way, way back on herbicide use. Kurt said these sprays also loosen soil structure and digest herbicide residues on the field.

"They got two different sprays they want you to spray either in the fall or in the spring, using chemicals from the oil slick cleaning-up industry that eat the hydrocarbons. It eats oil, you know. They dump it on the oil slick. And they spray it on the ground to eat all the chemical residue from herbicides and such. Their claims are that it helps to loosen up the soil and makes the soil so much more porous."

I didn't follow how reducing herbicide residue would loosen the soil, or how this would result in lower herbicide use. Kurt said he didn't know. But Kurt said that this is what their neighbor Rex has found. Kurt was also impressed that the company that makes these products is farmer-owned.

"It's a company that's solely owned by farmers. All the dealers are farmers. They've got some pretty radical products. But the end result is—"

"It's also got a radical structure, too," I added.

"Yeah, it is. It's a real radical group. And they're big. I mean, they're all over Canada, or parts of Canada. They're big in parts of the U.S."

I nodded, thinking it curious that I hadn't heard of them.

"They have test plots all over this area," Kurt continued. "But I could never get Gaylan to go to them. Well, Rex used that on all of their acres the last two years, and they're extremely happy. And Rex was talking about it last fall, bragging about how little herbicide they used last year. He said it was just incredible how little herbicide they bought. And Gaylan's standing there, and he's going, 'Well if it's such a good idea, why aren't the big companies promoting it?'"

"It's like, whew!" exclaimed Renea.

"He will not [accept it]. 'Nope, it can't work. It ain't going to work because they don't tell me that at the chemical dealer. And Fred used to work for DuPont, and Fred would know if it worked.'"

I have to say that these products didn't sound very plausible to me either, and still don't. In a recent check of the Internet, I couldn't even find the company name that Kurt had mentioned. But the point here is the way that Kurt located each side in the family dispute over these products within a different knowledge cultivation—within a different culture of knowledge and its social relations of identity. For Kurt, these sprays represented potential ways for farmers to pollute less and spend less. Moreover, they came from a relatively small company—even though he claimed it was "big" in Canada and the United States, which I think he thought helped validate the truth of the sprays' effectiveness. And because the company was small (but big enough to validate its success), it could be farmer-owned. Kurt thus understood the company's products to be farmers' own knowledge, not the knowledge of some "big company" out to make big bucks for its investors. These products were also the knowledge of the farmer next door, a farmer who, like Kurt, is interested in sustainable agriculture. So, for Kurt, these sprays were trustworthy.

Kurt represented Gaylan's knowledge cultivation, by contrast, as that of big industry and its more-is-better logic, happy to promote the pollution of the earth and the impoverishment of farmers as long as they can make money from it. He argued that this industrial truth is what Gaylan trusts, as shown by his remark that if these products were a good idea then the big companies would be making them. Kurt argued that Gaylan sought further validation from the chemical dealer and from "Fred"—not because Fred is a farmer but because he used to work for DuPont. So even Gaylan's local relations of knowledge weren't local, Kurt argued; they were part of the long arm of corporate capitalism.

I don't know what Gaylan himself would say about this. I was not able to interview him. My guess, though, is that Kurt was indeed right that he and Gaylan drew on very different social relations in assessing the trustworthiness of knowledge. I imagine that Gaylan would represent these differences, well, differently, but I think he would probably agree that the differences are there. I see no reason to doubt Kurt's word on that, much as I might doubt the effectiveness of those sprays.

Besides, I was doing the same thing that Kurt was in that interview: locating those sprays within social relations of knowledge. While I did feel some sense of trust in hearing that a local farmer had used the sprays successfully, I wanted to know what my friends in the agronomy department at Iowa State University would say about them. (They'd never heard of them, it turned out.) After all, I'm a professor myself, and I typically identify with academic knowledge—or at least with some of the many communities of academic knowledge. And although I didn't ask, I wanted to know if Kurt's neighbor, Rex, had maybe been approached by the company that makes the product about being a local dealer. (The arm of corporate capitalism is indeed long, and often disguised.) I wanted to know not only *what* knowledge, but also *whose* knowledge before I could trust it—and so did Kurt and Gaylan.

Like knowledge and the socially situated self, truth and trust are closely interrelated.[3] The cultivation of knowledge is as much about the cultivation of trust as it is about the cultivation of truth. A huge proportion of what we know we learn from others. It must be this way. Who has time to try everything for herself or himself? I am not a farmer, or an agronomist, so it would be very hard for me to try out these sprays—particularly on a big enough scale to really assess them carefully. It would take me years of training to become either a farmer or an agronomist. I must ask others, others I know and others whose own others I think I at least have a fair idea about: others I trust and others whose others I trust.

Which makes the phenomenological problem of farming even deeper. In order to farm, you need to plant some cultivar of your crop that has been developed by others over the decades, and even the centuries and millennia. If you're going to try to raise a corn crop this year, you can't start by going down to Mexico and collecting seeds from wild corn in order to develop your own cultivar first, or you probably won't get much of a crop. The same is true for everything a farmer does. In order to get through the day, a farmer needs *cultivars of knowledge*—lines of knowledge history, stretching through time and social relations—that is, recipes and routines of knowledge worked out through previous experience, and passed on to someone else for further working out. In other words, a farmer needs cultivars that come, at least in part, from others.

And so it is for all of us. For we know that every day is different, and we must hope that our cultivars will handle those differences, differences that we have not ourselves had the fortune or ill fortune yet to experience. So we borrow—and modify, to be sure—cultivars of knowledge that others suggest have worked across the differences that they have faced. Thus our whole order of being rests upon our trust in the truth of the order of other beings. We thereby become them, and they us. In this way, cultivars of knowledge intertwine with the cultivation of knowledge—with our cultural location of our identity within the social relations of knowledge. And in this way, the cultivation of knowledge becomes not only the cultivation of self but the cultivation of the social phenomenology of being.

And that's really a lot to give up.

"Where the heck is Donna?" I thought to myself as I sat in the dark in the Iowa State University car that I drove down to Baden, one of the smaller towns in Powell County. I was parked at Casey's gasoline and convenience store. I'd never been to Baden before. The only other retail establishment in town, as far as I could tell, was the exuberantly named Bill's Bait, Tackle, Archery, Ammo, and Supplies, just down the road. Donna and I had agreed to rendezvous at Casey's, an easy landmark, so we could both arrive at tonight's meeting in Donna's car, leaving my university one, with "Iowa State University" emblazoned on the door, covertly in the parking lot.

I reached for the tuning knob on the radio to continue my little survey of how many country music stations you can get in Powell County. (Ever the social scientist.) Someone knocked on the window. Donna.

"How is secret agent Mike this evening?" Donna said with a laugh as I stepped out of the car. It did feel a little like a spying operation. A farmer Donna interviewed had suggested we attend a meeting of "Frankie's group"—a kind of inspirational gathering for those interested in the truth about Hitler's Germany, in making biblical law the foundation of our system of justice, in stopping the coming of world government, and in escaping control by the federal government that is so bent on the destruction of farmers and local communities. "Patriots" is what they call themselves. A lot of farmers who have lost their land attend on a monthly basis, Donna's interviewee told her. But it was not the kind of gathering that Donna and I personally frequent. And we'd been advised not to be too obvious about our connections with the state university. So we were rather nervous about the whole business.

As I climbed into the front passenger seat in Donna's car, she moved a plate of homemade bars, wrapped in plastic wrap, into the back. "Eats? We were supposed to bring eats?" I was a bit alarmed and was thinking about dashing into Casey's for something.

Donna laughed again. "You're always supposed to bring something to a farmer's meeting. You'll see. There'll be a table full of stuff. But don't worry, Mike. You're a man. You don't need to bring anything. Just the women."

The group was gathering in the basement meeting room of Baden's grain elevator. Like many elevators in Iowa's small towns, this one is cooperatively owned by area farmers. Many Midwestern elevators have been bought up by grain conglomerates, but this one was still a co-op. And like most grain co-ops, it has a small room for membership meetings and local events.

We parked in the gravel lot, found the right door, and headed down the stairs, only a few minutes late. About twenty people, still in pre-meeting chatter, sat around several of those rectangular Formica folding tables that are stock items at cheap meeting facilities everywhere. Sure enough, one toward the back had several dishes of baked goods and snacks on it. Donna added hers to the collection. Then she gave me a little wave and went off to sit at what I suddenly realized was the women's table. Dumbfounded, I took a seat at one of the tables where the men were sitting.[4]

Most of the attendees seemed to be in their fifties and sixties, some older, a couple of people in their forties, and one that seemed to be about thirty. Most were men; there was just Donna and four other women. (And, come to think of it now, there were just about that many plates of baked goods and snacks.) Seated next to me was Floyd, one of the older men, a World War II veteran, he proudly told me. But Floyd wouldn't fight in World War III, if there were one today.

"Not with Bill Clinton as president." Clinton was still in office at the time, and the September 11th attacks were as yet beyond imagination. "Not in Bill Clinton's presidency. Because he just wants everything to be part of that one-world government. And I'd [wind up] fighting for the one-world government."

Eventually, a man in his mid-forties in a plaid shirt and jeans got up at the front of the room and began to speak extemporaneously. He talked about how he sees a lot of the country because of his job driving a truck. How things are happening everywhere. How people are getting together in their local communities and resisting the domination going on. How a black man in California was beaten to death in jail and how the authorities had told his family one lie after another about what had happened. How they and their community were fighting to force the truth to come out.

For a minute, I thought we'd come on the wrong night. The man with the plaid shirt could have been speaking at a labor rally or at a community organizers' meeting.

"We've got to have hope," he continued. "We've got to get together. People, things are happening. Things are happening in different parts of the country. We've got to get together."

But then he went on to describe how some communities are rejecting the "illegal legal system" of the United States, as he described it, and establishing their own local "common law courts" based on the Ten Commandments. How the government is controlling our lives. How taxes are "just another way they get you." How all U.S. lawyers are "foreign agents" because the American Bar Association was founded in England in 1788. (For what it's worth, the American Bar Association itself claims that it was founded in 1878 in Saratoga Springs, New York.)[5]

He passed some literature around the room—recent issues of the *Spotlight*, the *Prophecy Club Newsletter*, the *Bob Livingston Letter*, and *Scriptures for America*. An article in *Scriptures for America* caught my eye: "Concerning the Oklahoma Bombing: Solving the *Who Done It* Mystery." (Jews, the article argued.) The *Prophecy Club Newsletter*, it turned out, is a kind of tour schedule for the far-right national speakers circuit. The copy that I was given listed the Holiday Inns, Marriotts, Hiltons, and Ramada Inns in twenty-one states across the West and Midwest where Eric Barger would be speaking on "It's the International Bankers—Not the Democrats and Republicans," where Dr. Mike Coffman would be addressing the subject of "Environmentalism—Door to World Religion and Government," and where Ron Wyatt would be showing a video of his discovery of the ark of the covenant on Mt. Sinai, which, the newsletter says, God had called him to find.

After about half an hour, an older man in shirt and tie got up to speak. This, as we learned, was Frankie, the leader of the group. Frankie went on for nearly another hour, covering a wide range of themes, often jumping suddenly from topic to topic. He spoke at length about Nazi Germany, claiming that Hitler is misunderstood and was actually a great leader who reinvigorated the German economy. Most of the evidence for the Holocaust, he said, was made up by Jews and others who don't want us to know the truth about how we could have a government that worked for the people. In America, "the government has stolen the national wealth and called it the national debt." And "in two or three years, the government is going to bring us back to one-dollar grain. And basically going to try to wipe us all out." (Indeed, grain has often sold in that price range in the years since this meeting took place, contributing to many more farmers' leaving the land.) Another of his big themes was Ruby Ridge and the militia movement. Frankie agreed with their aims but not their approach. "Militias are making a mistake. Keep your gun, but keep it hid. Don't let your enemy know who you are and where you are." Don't be a militia man, he advised, be a "patriot." But recognize what those people out there in Montana (not Idaho, inexplicably) were doing: banding together to try to hold on to their land. "This is what the government is always trying to do. Take our land."

For most of the speech, no one said a word. The audience seemed passive and listless, despite the fire in Frankie's words. But when Frankie brought up the subject of government theft of people's land, Floyd turned around and said to a man next to him, evidently another farmer who had been forced out of farming, "Well, we know all about that, don't we?"

Militias, patriot groups, and other far-right organizations in fact have a surprising amount in common with the left. The devastation of local communities wrought by economic dislocation is also a devastation of selves and their phenomenologies of the world—their knowledge cultivations. Looking around the room, I got the impression that I was in the presence of the deeply wounded. Of farmers without farms. Of men deprived of much of the means to masculinity as they understand it. Of people whose principal source of connection with their heritage (again, as they understand it) has been ripped away from them. Of new residents of what Osha Gray Davidson terms the "rural ghetto."[6] "When all is said and done," notes the anthropologist Kathryn Marie Dudley, "the loss of a farm is not just the loss of material possessions: it is also the loss of a sense of community and one's place in the world."[7] Experiencing these deeper forms of "dispossession," as Dudley terms it, and a consequent sense of disorientation, former farmers seek a phenomenological lifeline. Patriot groups try to provide this secure connection to a new knowledge cultivation that people can identify with and use to reconstruct their everyday lives, one that is relevant to their experiences. And, as Davidson notes, "if genuine alternatives are not provided, a significant number of rural ghetto residents—bitter, desperate, and increasingly cut off from the nation's cities—are sure to seek their salvation in the politics of hate."[8]

I think Frankie understands this, at least to some extent. He knows he is trying to cultivate his listeners into different social relations of knowledge, identifying some knowledge as "our" knowledge and other knowledge as not, not at all. His is a cultivation with hard social boundaries, and the world of experience it describes is similarly hard-edged. But the theme he returned to most in his long, rambling speech was what he termed the "cognitive dissonance propaganda" of the government. As a social scientist, I was startled by his use of the term "cognitive dissonance," one of the principal concepts of social psychology. Frankie was evidently concerned that his audience wouldn't know these words and held up a little cardboard sign with "cognitive dissonance propaganda" written on it. "What do these words mean?" he asked the audience. No one answered, as was typical throughout the evening. "Brainwashing! When the truth comes out, you reject it."[9]

Some on the left might use the term "hegemony"—the idea that the rich and powerful control what everybody else thinks through manipulating the media and the educational system, and through other means.[10] But with either the term "hegemony" or Frankie's "cognitive dissonance propaganda," knowledge cultivations are equally identified as social threats to the cultivation the listener is thus expected to identify with. By identifying knowledge threats, the speaker encourages the hardening of the social boundaries of knowledge, building around a cultivation fences that keep people and knowledge in as much as they keep people and knowledge out.

The right and the left meet up in other ways on the back side of the ideological circle, usually unintentionally. Or so I found myself musing that night. One of Frankie's consistent themes was money—how it was "the root of all evil," how bankers control everything, how legal tender laws forced everyone to give up gold and silver and accept valueless paper, and how the new design of American money was just the U.S. government trying to prepare us for the coming of world government by making our bills look more European.[11] Money was behind everything, a materialist philosophy that had some passing similarity to simplistic versions of Marxism.

Frankie even brought up an increasingly popular cause of the left, the local currency movement. Community groups in several cities across the country have introduced currencies that are accepted only within the local region, thus encouraging capital to stay local and promote the local economy. The program in Ithaca, New York, is the most developed. I had just returned from Ithaca, where a good friend was working at the organic food co-op. He had been telling me that he now takes almost half his income in "Ithaca Hours," as the local currency there is called. It's good lefty stuff, but there was Frankie going on about it, waving around *New Money for Healthy Communities,* a book that describes the local currency experiments in some forty U.S. cities. Frankie's take on local currencies is a little different from most leftists', I think. He thought one of its biggest advantages was that maybe you wouldn't have to pay income tax. But even that is something some on the left might relish.

After Frankie finished, there was a break. People chatted and enjoyed the eats. I just had to have a look at the local currencies book, though. It turned out to have blurbs on the back from Wendell Berry, Ralph Borsodi, and Frederick Soddy, all of whom are often read by those on the environmental and communitarian left. Not surprisingly, Frankie hadn't mentioned this possible political association. Instead, he sought to locate socially the idea of local currency within the knowledge cultivation of patriot farmers that he was trying to develop through these monthly meetings.

Floyd saw me taking an interest in the book and asked why. So I explained that I had recently been to Ithaca, New York, and had seen this new money. Someone else overheard our conversation, and called over to Frankie and mentioned it to him. "Oh really?" he said, and started walking over. My heart sank. The whole evening, I had been feeling incredibly obvious as a left-wing infiltrator—which, given my own political views, I suppose I was. And I was sure that Frankie had earlier been giving me suspicious looks.

But, to my amazement, Frankie asked me mildly if, after the break, I would be willing to address the group about local currency, since I had actually seen the stuff. About the last thing I expected at a patriots' meeting was to be invited to give a lecture. Reluctantly, I agreed. (Ever the social scientist again.) It was the hardest lecture I've ever given: five minutes of sheer terror.

We bolted shortly afterward. But as we went, what Frankie earlier called his favorite Bible quotation was ringing in my ears: "My people perish for lack of knowledge." Hosea, chapter 4, verse 6.

Few of the Iowa farmers we interviewed identify with patriot groups and their knowledge cultivation. It is probably more attractive to former farmers, but I believe it is uncommon among them as well. Less extreme ideas along the same general lines are indeed common in rural Iowa, as they are throughout the United States, but they are enmeshed in different networks of social relations. Patriot groups aren't exactly packing the house in elevator basement after elevator basement—particularly since September 11th.

Still, Frankie's group's interest in local currency helps illuminate the way knowledge moves across the boundaries of what I've been calling knowledge cultivations. Different knowledge cultivations may draw on what were originally the same cultivars of knowledge and subsequently develop them in substantially different ways, recultivating their social relations and blending them with other cultivars. The local currency movement that Frankie has taken an interest in is an example of a cultivar that is being recultivated into a cultivation to which it is new. This cultivar may be quite different after Frankie and other "patriots" are done recultivating it. But their cultivation of the local currency cultivar will nonetheless retain a potential for connecting to the cultivation they got it from. That history is still there. And even if it weren't, connections between cultivations can always be found in the future (often through recultivating the past). In this we may find some hope for connecting people and their knowledges across the boundaries of cultivation.

But the crossing of the boundaries of cultivation is no easy matter, for the cultivation of knowledge is as much a matter of the cultivation of *unknowledge*—a matter of the identification of the knowledge of other social relations as inadequate, mistaken, or unimportant.[12] By unknowledge, I don't mean ignorance. To equate the two is immediately to choose sides and assert the hierarchy of knowledge that we all, I suspect, to some extent feel about what others think. It doesn't help us to linger here on that old point of social division. Still, we must all confess that getting through the day depends heavily on avoiding most of the cascade of knowledge that is continually available to us. As William James observed, "the human mind is essentially partial. It can be efficient at all only by *picking out* what to attend to, and ignoring everything else—by narrowing its point of view. Otherwise, what little strength it has is dispersed, and it loses its way altogether."[13]

In other words, farmers can't spend all day reading farming magazines, cruising the Internet, talking to sales people, and having coffee with locals at the café in town. These possibilities are available to them, but they, like the rest of us, must at some point get on with what needs getting on with (to paraphrase Garrison Keillor once again). We rely on our knowledge cultivations in part to identify what is not knowledge, and therefore not worth attending to, as much as to identify what we do need to attend to. Our knowledge cultivations tell us what to trust and what not to trust, and therefore what to disregard. Rather than the cultivation of ignorance, we should think of it as the cultivation of the *ignorable*.

But, of course, we do usually choose sides. And while that promotes divisions of social relations and their knowledges, it also helps maintain the larger phenomenology of human lives and ward off the threats posed by the existence of other cultivations. For other cultivations are indeed often quite threatening.[14]

One of the places where that sense of threat came out most strongly in our interviews with farmers was over environmentalism and the animal welfare and animal rights movement. Many farmers seemed profoundly unsettled by these public critiques of what they do and what they think, and seemed to take them as critiques of something deeper: who they are.

Gene's in his late thirties. He and his wife JoDee milk about eighty cows on their 250-acre farm. It's the "home place" for their family, and Gene was raised there. They bought it from Gene's dad a few years ago. His grandparents owned it before that. It's not a century farm, but it's pretty close to it. Farming is central to Gene's understanding of himself, and he told me that he never thought about doing anything else.

The setup on the farm is not the most up to date, but Gene and JoDee are making it work reasonably well. They don't have everything they want; still, as Gene said, "I guess we ain't really hurting." The main problem is the milking operation, Gene explained. Their dairy uses a stanchion system in which the whole herd of cows lines up at feeding troughs at milking time and the person doing the milking has to move the "cluster"—a kind of mechanical hand with four suction hoses, one for each teat—from cow to cow. Many U.S. dairy farms over the past thirty years have switched to milking "parlors" in which the clusters stay put and the cows do the moving, which most farmers find considerably more efficient. Stanchion barns are also often criticized as being less humane, because a cow is generally restrained with a frame around her neck, or with a chain, for far longer than is needed for the actual milking. Some farmers leave their cows chained in stanchions for most of the day. Also, older stanchion barns were built with a smaller cow in mind than the product of today's cow breeding. The stalls are simply too small now in many stanchion barns, and the cows find them quite uncomfortable, particularly if the stanchions still operate with frames, which are more restrictive.

Some stanchion barns have been modified to work through many of these problems, however, and I wanted to see theirs before our interview began. But Gene was reluctant, so I didn't push it. I early on got the sense that Gene has strong feelings about being criticized by others about what he does. We were talking about sustainable agriculture. It wasn't a phrase he'd heard before, and in this he was no different from about 40 percent of Iowa farmers, according to a 1998 survey. (When asked what sustainable agriculture meant, 47 percent of the roughly one thousand farmers in that survey offered no opinion or said that they didn't know.)[15] And we talked about environmental issues like groundwater pollution. He got, frankly, a bit testy.

"Well, I don't put on any more spray, or anything like that, than I feel I absolutely have to. I don't waste any of that. I mean, if it ran down into the stream it's because God sent too much fricken rain that night or something. It ain't because I put too much on."

He was on edge for most of the interview, and I think he was worried that I might be an environmentalist or animal welfare advocate myself. Which I am (although I have some doubts about animal rights, as opposed to animal welfare). I hadn't told him any of that, but I think he was put on his guard by some of the questions I was asking. It was also one of the first interviews I had done for the book, and I was a bit nervous about it myself. Plus I had gotten his name from a neighboring farmer whom I'd met at a Practical Farmers of Iowa field day. So he probably associated me in his mind with his neighbor, who is quite concerned about environmental issues.

But I think I won him over a bit by the end of the interview. He won me over somewhat, too. Whatever our private suspicions, I think we discovered that we genuinely liked each other.

After about an hour, we got to talking about social class and where he sees himself with respect to that. I was always a bit worried asking farmers about class, as Americans often find it a sensitive topic. But Gene really opened up at this point.

"I think of myself as being up there," he said, not in an arrogant way. "But I think everybody else thinks that farmers are farmers because that's all they know."

"Right," I nodded.

"I have a higher opinion of myself than that," he went on with a laugh.

"Right, right. Right." I agree that there are some negative images of farmers in contemporary culture. But I also think they jostle alongside many positive ones. "Well," I began, considering whether to bring this up, "some people see farming as a kind of a—you know, you get this image of farming from children's books, and sometimes from movies and things, that farming's almost kind of a noble sort of profession. There's quite a bit of romance in it, if you know what I mean."

"Yeah," Gene replied, but I don't think he agreed with me. Or rather, he agreed in a different way. "I kind of feel that way too."

"You do?" That was a pretty stupid thing for an interviewer to say. But Gene didn't take offense. I think he took my line of questioning as a sign that I respect farmers, despite the awkwardness of that "You do?" In any event, he immediately went on to tell a story that underscored his feeling that he personally finds farming noble and romantic, even though, in his view, the general public doesn't.

"We were at a meeting of our dairy cooperative, and they had all these different speakers. And one of them, she was a vegan lady—an animal rights activist."

"Sure. Vegan. They won't eat anything [from animals]. They won't wear leather shoes, or—" I was trying to show that I thought vegans were extreme. My views on veganism, and vegetarianism more generally, are actually pretty complex. But this wasn't the time for that. Besides, he cut me off in mid-sentence.

"She had no use for us even milking cows," Gene continued. He laughed, rather uproariously, at the thought. I joined his laugh. "She didn't. I mean, she got into how taking that calf from that mother is the absolute worst thing that can ever happen."

"That's what she said?"

"More or less. Now woman, you're in a different land than what we're in. I mean, we make our living how good we treat our animals. The better we treat them, you know, the more we're going to make. Then she accused us of everything we do for animals is because of economics. Well, I wonder if her animal rights people were paying her at all. She was gaining by the animal also, if it come right down to it." He gave a kind of smirking laugh, trying to be light while still making this pointed accusation.

"Uh-huh, right," I said, "uh-huh."

"I don't know. I don't think I mistreat any animals. But she, just by the mere fact that we milked them, it was mistreating them." He laughed again.

"Well, if you weren't milking those animals, they wouldn't even be alive," I said, trying to be supportive. "They wouldn't even have a life."

"Yeah." Gene nodded emphatically, and went on to raise a related point. "Well, one person asked her what would they be on the earth for. And she compared them to deer. They should be left to roam like deer. Automatically, there wouldn't be so many. I don't know how my cows would've fared out for two weeks or a month." (I think he meant "fared outside on their own.")

"So she'd rather have the cows be eaten by wolves?"

"Well, yeah, that would be the natural—"

"That would be less cruel than what you do," I suggested, meaning that would be her view.

Gene nodded. "That would be the natural thing," he then repeated with a laugh, this time finishing the statement. "Yeah. I don't know. She'd sooner see us hit them with cars and all that on the highway. I don't know where she was coming from."

"Well, would you say that people in general respect farmers? Say the public does?"

"I think around here they do. Yeah. I think people like her have absolutely no respect for us at all. They think we're hicks. I'm sure she thinks we're all hicks. And I bet there's a lot of people that have no idea of what farming is about that think that too. And now with all these—"

Environmentalists, I think he was about to say. He didn't finish the sentence, but what he went on to say made it plain enough. "There's chemical and stuff getting in the water. I'm sure there's people think farmers are ruining our water, you know. Well, we're doing everything we can to be as good about it as what we can. We still gotta produce. They don't realize that everything they get in the grocery is coming from us."

Gene's ire is not only raised by veganism, in other words. Among the "lot of people" who don't understand farming are environmentalists in the cities, and maybe city people in general.

"Well, it's not like people who live in cities don't pollute too," I said, trying to be supportive again. I may have been leading him a bit here on the subject of city people, but he picked up on it readily enough.

"Yeah," Gene went on. "We had to apply for a pesticide license. That was around fifteen bucks a hit from us, that they have no business getting either." Another laugh. "Someone asked about people fertilizing their lawn." I think that by "someone" Gene meant another farmer. "They use way more per acre on their lawns then what we do out here on the farm. But it's us farmers that are polluting everything."

We joined in yet another knowing and ambivalent laugh.

"But no, I've lost that question again too. Forgot it."

"Oh, well, we were off on other stuff," I said. "That's all right."

"Was it how other people look at farmers or something?"

"Right. Do other people respect farmers?"

"I think in this area we're all mostly farmers," Gene replied, meaning that therefore local people respect farmers. "People in the small communities are making their living off of farmers. They better show a little respect."

And once again that laugh, that ambivalent laugh.

After the interview, Gene did show me the stanchion barn. They were no longer using frames around the cows' necks, at least. They had installed chains, which give today's big cows more freedom of movement. Still, it seemed to me a pretty dark and uncomfortable place to be a cow. I hope he gets the money to update it soon. But it matched our conversation that day, which also had gone into some dark and uncomfortable places.

For part of Gene's strategy for dealing with the threat posed by environmental and animal welfare critics is through identifying their critiques within particular social relations of knowledge and through arguing that those social relations are far from disinterested. The "vegan lady" is an "animal rights activist." She lives "in a different land" conceptually, and her social relations are those of "her animal rights people." Moreover, these animal rights people were probably paying her, so that "she was gaining by the animal also." And she doesn't respect farmers; Gene's sure "she thinks we're all hicks."

In other words, the identification of the social relations of knowledge is not something one does only with respect to one's own knowledge cultivation. It is also something that one does with the cultivations of others, but generally with a rather different intent: to point out power relations. The "vegan lady" is paid to say what she says, Gene suggests. By emphasizing the femaleness of the animal rights advocate with phrases like "vegan lady," Gene may also have been trying to introduce a note of gender conflict into his counter-critique, of female squeamishness threatening the livelihood of the farming man. He did not explicitly claim that the animal rights advocate was trying to advance women's power over men through her arguments.[16] (Among other things, I think that would be a tough claim to establish.) But through pointing out and hinting at the dark and uncomfortable material implications of knowledge cultivations, Gene sought to undermine the justice of the animal rights and environmental critiques.

Material criticism is a powerful line of argument, pun intended, for the social relations of knowledge are indeed as well the power relations of knowledge. To speak of knowledge is to speak of identity, and vice versa. But to speak of identity is also to speak of power, and vice versa. And to speak of power is to speak of knowledge, and vice versa. It's circles within circles, with each element of equal significance, equally influencing the other. The French philosopher Michel Foucault wrote widely about what he called "power/knowledge."[17] He was two-thirds right: Although it's awkward to say, it's really *power/knowledge/identity.*

Central to the circles of power/knowledge/identity, of the cultivation of knowledge and unknowledge, is the *naming of knowledge*—often literally. The *Monroe* doctrine. Pico*curies* and mega*hertz*. *Marx*ism. The academic's footnotes and lists of references. Similarly, Gene, Frankie, Kurt, Renea, and (albeit secondhand) Gaylan continually sought to understand knowledge's social relations by associating it with names—with the names of social groups, and often with the names of individual people. There is the knowledge of farmers, of your neighbor Rex, of corporations, of your friend who used to work at Dupont, of public figures like Alan Savory, of university scientists, of the government, of the militia movement, of "patriots," of Christians, of animal rights activists, of environmentalists, of your own experience. I've been doing it too, most recently in the paragraph above where I cite Michel Foucault. And in a way, I do it every time I bring another Iowa farmer into this book.

Understanding the truth is all about evidence, the scientist might say. The question is, evidence of what? And to answer that we must also ask, evidence of whom? By locating evidence in the social book of names, we establish relations of trust and truth. We name the phenomena of our lives, and thereby grasp our own phenomenology. By naming knowledge, we name ourselves.

Not all cultivations are the same, however. The naming of knowledge, the identification of interest, the cultivation of knowledge, and the cultivation of the ignorable—these processes are common to cultivation. But some cultivations constitute themselves through more monologic relations of self and other, and some cultivations through more dialogic relations. By this distinction I mean the way that some cultivations are more inviting to the critiques of those outside them, and use the identification of interest as a means of understanding and engaging difference, not rejecting it. Perhaps this only shows where my own sympathies lie. Nevertheless, I could not help a feeling that Gene sought a kind of monological comfort in using the identification of interest to reject what the "vegan lady" had to say, instead of using it to understand what she had to say.

Gnothi seauton, "know thyself," was what the ancient Greeks carved above the entrance to the Oracle of Apollo at Delphi, and it is still good advice, for farmers and everyone else. And the way we know ourselves is to self our knowledge, and by identifying other selves with other knowings. These are not always happy processes, and they lead as much to unknowledge as to knowledge. To farm the self is to cultivate our allegiances and to align our practices along the rows of a social agronomy. Not that we could, or should, do otherwise. But we could do so in ways that tend toward the monological, or in ways that tend toward the dialogical. In either case, these patterned inclusions and exclusions are structures of a sort, as influential on the lives and ways of farmers as government, technology, and the economy. These are *all* structures of agriculture.

Intermezzo

One of the conventional distinctions in social thought is between structure and culture. This distinction is typically also associated with the view that structure is the serious stuff of social life, versus the ineffable and unstable that is culture.[1] Structure is all those constraints you can't do much about, at least on your own, because they are so much bigger than you, came before you, and will be there after you—things like the governmental, economic, and technological factors referred to in the conventional meaning of the phrase the "structure of agriculture." In contrast, culture is values, beliefs, disbeliefs, desires, and favored ways for getting through the everyday.[2] Such a view leads immediately to a paradox. On the one hand it leads to a micro view of culture as something local and small, and thus something over which we have some considerable freedom of choice. You can always change your values if you don't like them, right? But that's just because culture is inconsequential. On the other hand, it leads to the view of culture as an epiphenomenon, as taking its shape in the main from the shape of structure and as having little independent power of its own. Where do your values, beliefs, disbeliefs, desires, and daily practices come from to begin with? From what you have to do to make a living. From the laws you must follow. From the technologies of your life. So you don't have much choice in your culture, because culture comes from that which you cannot control: structure.[3]

But culture is neither inconsequential nor an epiphenomenon. Although it can be found at the level of the small and local, culture too is bigger than any one of us, was here before us, and will be here after us. The processes of the cultivation of knowledge that I describe in the previous two chapters extend well beyond the local realms we find ourselves in. They depend on knowledge cultivars from outside those realms, and they connect with people far away, people we have never met and never will meet. These processes also involve various forms of identification and dis-identification with governmental, economic, and technological forces. In other words, the cultivation of knowledge is intimately bound up with the factors conventionally considered under the notion of the structure of agriculture. But that does not make knowledge cultivation an inconsequential or epiphenomenal product of those factors. Given how ill treated most Iowa farms are by those factors, without the cultivation of knowledge—without the social relations of identification, knowledge, and (as will become clearer in the next chapter) power—it is unlikely that those factors would persist without some additional, and extraordinary, application of external coercion.

In other words, culture too is part of the serious stuff of social life. If that seriousness—if a patterning persistence across social space and social time—is what we mean by structure, then culture too must be seen as structural in a way, as the social theorist Anthony Giddens has argued (or has as good as argued). Strictly speaking, Giddens's definition of structure is: "the rules and resources, or sets of transformation relations, organised as properties of social systems—the structuring properties allowing the binding of time and space in social systems."[4] But basically what he means by structure is social things that hang on across space and time and pattern what we do and don't do, whatever those social things may be. Giddens doesn't have a lot of use for the word culture, however. For him, the matters we have conventionally referred to as culture and those we have conventionally referred to as structure are all potentially binders in this "binding of space and time." For him, it doesn't help much to use the words culture and structure to keep track of some inevitably arbitrary distinction among the kinds of binders. So he calls it all structure, and talks about how these binders "structurate" social life.

The theorist Pierre Bourdieu made a similar point. He argued that sociology needs to focus on tracing the connections and interactions between "objective structures" and "mental structures," thereby overcoming the field's traditional opposition of the objective and the subjective. For Bourdieu, then, culture was something to be explained through these structured connections, but culture was of limited analytic help in doing the explaining.[5] The theorist Margaret Archer disagrees, however, albeit not completely. She argues that Giddens's approach runs too much together and that the distinction between culture and structure is analytically very important. For Archer, structure refers to organizational and material matters, while culture refers to "intelligibilia," the realm of our ideas and understandings. She thinks it's clearer to say that both culture and structure are matters that hang on and pattern what we do and don't do.[6]

Why am I going into this what-is-culture, what-is-structure debate? Because I would put it all still differently again, and yet also very much the same, and I have organized the argument of this book along these similar-yet-different lines. The first section of the book enters what I see as *ecological dialogue* from a materialist moment in this dialogue, and the second section from an idealist moment. Matters of economy, state power, technology, and the like are, in the first instance of our imagination of them, material factors; and matters of identity, knowledge, social affiliation, and the like are ideal factors in our first imagination of them.

Both sorts of matters often persist over social space and social time, patterning what we do and don't do, and in that sense they are equally structural. Yet that second instance is never far behind, for material factors depend upon ideal factors and ideal factors depend upon material factors to attain their persistence. The economy of agriculture embodies cultural values like the virtue of competitive individualism as much as it undermines cultural values like the importance of communal ties. The cultivation of knowledge depends in part upon our sense of the material implications of that knowledge as much as it leads to the persistence of those material implications. The attraction many farmers feel for Big Ag depends in part on their sense that it will lead to a big harvest of grain and gain, even if it leads only to acceleration of the treadmill of the farmer's problem. In actuality, there is no first instance or second instance of either the material or the ideal. They are an endless conversation, each mutually shaping the other. But much as we take turns speaking in a verbal conversation, so we must take analytic turns in the material-ideal voices of ecological dialogue, if we are to understand its dynamic and the contributions of its constituents.

Yet these are not matters only of constraint, as Giddens, Bourdieu, and Archer all agree in their own ways. Sociology typically dwells on social constraints, however, and thereby risks becoming its own kind of dismal science. Thus far in this book, I too have dwelled on social constraints, in both their material and ideal dimensions. But my purpose in doing so has been to better understand the wonder of how Iowa farmers sometimes break free of them. To that, in my view, equally serious but happier topic I now turn.

PART III

THE SUSTAINABLE LANDSCAPE OF PRACTICAL AGRICULTURE

6. Rolling a New Cob

"I probably played the board for, oh, six to eight years—and lost money all the time. Some years it wasn't a lot. But instead of making that extra ten thousand a year, you end up losing the ten thousand. If you do that too many years in a row, why, you just can't overcome that."

Dale and I are sitting in the comfortable kitchen of the farm where he and his wife, Yvonne, and their two high-school-age children raise corn, soybeans, goats, and hogs, using many of the techniques that sustainable agriculturalists advocate. They do ridge tilling, like Dick Thompson does, to minimize herbicides. They rotate their crops and pastures over a five-to-seven-year period to break up pest cycles and conserve soil. They use rotational grazing to increase grass productivity. They farrow their pigs outside on their pastures, giving the pigs the opportunity for fresh air and thereby lowering their costs for antibiotics and farm buildings. They feed most of their grain to their own livestock, so Dale no longer has to "play the board"—that's the Chicago Board of Trade, the fickle finger of financial fate for grain farms across the United States—to make sure he's getting every penny out of their corn and soybeans. It's a diverse operation that spreads their economic risk out over several products. And they're even talking about starting a goat dairy. Dale and Yvonne's is not your typical Powell County farm.

At least not anymore. As the kids play some noisy video game in the den, and while Yvonne is out checking on the goats, Dale is telling me about how a few years ago they radically rewrote the script for their farm.

"I'd have to say that I'm a pretty fair market technician. When you've got all these experts together and all that market bullshit, well, you know, I can do all that too. I'm a student of the markets," Dale says with some pride. He thinks over the past. "What I spent a lot of my time on was mainly marketing, not chemicals or production. It was mainly marketing, and studying it, and then doing all the technical stuff every day."

He clears his throat. This isn't easy. "And the cycles were such that we should have had a drought and then a grain rally three years ago. I kept playing the adage that this was going to be the drought year. So I wanted to be prepared myself. And it didn't happen, and it didn't happen, and it didn't happen."

Gary Guthrie (right), PFI *member and* CSA *farmer, with his son, Erik, at their farm near Nevada, Iowa, 1998.*

The kids shout something. Their game is going well. We exchange grins. Then Dale's tone turns rueful as he goes back to his story.

"So you know, well, I, you feel that your timing wasn't right, you know, because if things would have happened—" as one planned, he implies, letting his sentence hang in the air. "If it wouldn't have rained like one weekend here, oh, a couple three years ago when I was playing it. But it happened to rain that one weekend. We was really shaping up for a drought. You know, I had the boat loaded. But it rained over the weekend, so that was the end of that deal." Dale laughs. "The boat sunk in the harbor!" Dale laughs again, harder.

That's why they call it "playing" the board. It's a crapshoot. You do your best, but it's still a game of chance. Dale held his grain back from the market, anticipating a drought that would send prices for grain futures higher. He had crafted a plan to be on the upside of the farmer's problem—to have lots to sell at a moment when most farmers had little. But then it rained, prices dropped, and that extra $10,000 turned into another $10,000 loss.

Dale was devastated. Here he was doing just what the conventional experts say farmers should do: farm the market, not the farm. In this day and age of industrialization of production, the farming part should be a breeze—a recipe, a transportable production line that you buy from the chemical and implement dealers and set up in your fields. We've got that part covered for you, the agricultural corporations say. We're getting you off your tractor seat so you can get in front of the computer screen and watch the market reports stream in.

But it didn't work for Dale. Time and again it didn't work, even though he was an active participant in a "marketing club"—a group of local farmers who got together regularly to discuss the obscure ways of the Board of Trade. Finally, it all seemed to come down to that weekend three years ago when he found himself in an odd position for a farmer: praying for more bad weather.

"But you know, going back, all these chains of events that have happened in my life, I'd have to say that I've been a better person because of it. And I have no remorse for the money I've lost or whatever. Because, the thing is, to me, it made me a better person. And I had to go through that. It was almost like it was a planned event. That in order for me to understand things and the way they are today and then maybe to have the goal and—"

"When you say planned, you mean by something higher?" I interject, perhaps with a bit of a sociological pounce.

Dale looks at me carefully. This isn't the kind of thing men talk about among themselves very often. But it's something he wanted very much to say. "Right, yeah," he confirms, choking back a bit. On the tape you can hear his voice dropping in a slow "ri-i-i-ight" as he looks straight into my eyes.[1]

What Dale was talking about in his kitchen late that afternoon was something many of the sustainable farmers we talked with described. Ours wasn't the kind of research calculated to produce clear numbers for measuring the weights and proportions of social life. But I think it significant that more than half the farmers we were able to speak to in detail about how they came to sustainable agriculture reported a similar experience: a sudden, disorienting change, in most cases during a period of severe economic stress, in which they had to rethink not only their farming practices but their practices of self. Dale's description of it as something "I had to go through." The way he can pinpoint a specific event that finally brought it about. Dale's feeling that "I've been a better person because of it." All of this speaks to a sudden rip in an overstretched fabric. It was a moment that ruptured his phenomenology of farming, his sense of what and who could be trusted to hold together. Sometimes you can repair old fabric if you get a patch on it quick enough. But sometimes the rip is so deep and so sudden, especially when the cloth is weakened to begin with, that a patch large enough and strong enough isn't on hand to prevent the shredding of the everyday in the flapping of the breeze.

We usually understand economics as the realm of money and necessity—as a matter of "dull compulsion," in the words of that trenchant observer of the life of capitalism, Karl Marx. But economics is far more than money, and far more than necessity. It is about what we can and cannot do, what we know and do not know to do, who we are and who we are not, not just in the gaining of our living but in the *living* of our living. It is part of our phenomenology of being and doing. Economics is a structure that gives shape to our daily practices, just as our daily practices in turn give shape to that structure. Corporations, commodity cycles, and the Chicago Board of Trade are central structures in Dale's daily practices. So too are the social boundaries on his knowledge of the ways of corporations, commodity cycles, and boards of trade, boundaries that go well beyond what a local marketing club can transcend. Two children to raise, a lifelong connection to Powell County, and rain that falls at the wrong time are equally economic structures that pattern what Dale finds himself doing. To see economics as merely about money and its compulsions is to miss its phenomenological dimension. Economics, in the sense I am trying to suggest here, is part of *the structure of practice and the practice of structure.* It is not just in the breeze but also in the fabric.

The interrelationship between structure and practice is not, however, a deterministic one. The economic and other compulsions we sense and therefore practice sometimes rip apart, as they did for Dale. We do not necessarily feel such a rip as the freedom it could be, though, at least not immediately. While it may shred our sense of constraint, of what we feel we have to do, it as well shreds our sense of constitution, of what we feel we know about the world and therefore should do in the world. We doubt what I have been calling our knowledge cultivations.

And we doubt the social relations that underpin those cultivations. In the recognition of economic structures and practices that no longer seem to work for you comes, just as surely, the parallel recognition that they are working for somebody else. No longer an uncontrolled breeze testing a neutral fabric, the economic suddenly appears as a "guided doing," to use the apt words of the social theorist Erving Goffman.[2] There are interests in the blowing of the economic wind and in the recommended weaves of living, social relations of interest that we now recognize to be not our own. Thus, not only is our sense of knowledge left gaping and frayed, but also our sense of trust in others.

We may even come to doubt our own selves, as Dale describes it.

"I kind of lost confidence in myself, because some of the things I was doing before failed me. Naturally after your ideas and things had failed, well, then, soon you get kind of gun shy. It's taken me probably a couple years to gain my confidence back, so to speak. You know, making sure that my decisions are based on sound decisions—based on goals and beliefs rather than based on paradigms, peer pressure, and prejudices."

"Your three P's? Right?"

He laughed a laugh of yes. Dale and I had met up several times before our interview in his kitchen. We all have our little raps, and one of Dale's is his rejection of what he calls the "three P's" of paradigms, peer pressure, and prejudices. He mentioned it so often it had become a little joke between us.

It's a good little rap, very good indeed. I call it knowledge cultivation, but in most respects Dale's three P's make the same connections. By linking paradigms to peers, Dale points out how knowledge is cultivated by cultures and how cultures are cultivated by knowledge. By linking them both to prejudices, he points out how knowledge cultivations commonly have an exclusionary dimension, excluding both people and the particular currents of thought they bear and that bear them along; he is talking about the cultivation of unknowledge, the waterfalls of the ignorable that we trust our knowledge cultivations to identify safely for us. He is talking about the cultivation of monologue.

Dale is also pointing to the external origin of these paradigms, this peer pressure, this prejudice. In his description of them, he made it clear that the three P's come as much from the outside as the inside, if not more so. The paradigms that are collectively held. The pressure that is placed on one. The prejudice that he now sees as unjust. The three P's are the three P's of interest as much as of social relations, for interests and social relations have common origins. Dale's three P's thus are his way to describe what a struggle it can be to recognize monologue.

But there is an important difference between Dale's rap and mine. Dale believes he has escaped the three P's. I'm not so sanguine. Sustainable farmers certainly use conceptual orientations they have learned, at least in part, from others. They certainly depend on and value the views of other sustainable farmers and of others committed to sustainable agriculture. There certainly is a strong element of commitment among sustainable agriculture people, and a sense of skepticism about some of the ideas and motives of nonsustainable agriculture people. And there are indeed interests behind these orientations, commitments, and skepticisms. That sounds like the three P's to me—different paradigms, peers, prejudices, and interests, to be sure, but paradigms, peers, prejudices, and interests all the same.

There's an old expression you still sometimes hear in the rural Midwest: "rolling the cob." It means having an unhurried good chat with friends and neighbors about the ways and doings of the world. It means talking across the fence with the farmer next door, idly rolling an old corn cob back and forth in the dust with your toe as you stand there. To roll the cob is to cultivate knowledge with others. And sustainable farmers (like all farmers, indeed, like all of us) still roll the cob.

I don't mean this as a criticism of sustainable agriculture, or of other approaches to agriculture, for that matter. I'm not sure that anyone can escape the cultivation of knowledge and its origins in patterns of social interest. For better or worse, knowledge is a social process. It could not be otherwise: humans are social.

But sustainable farmers are rolling a *new* cob, one I believe they have good reason to value. That cob is striking in its general embrace of more dialogic relations of self and other, as I'll describe in more detail in the following two chapters. (*More* dialogic: A degree of monologue is, I fear, inescapable.)

But I am getting ahead of the story. In this chapter, my focus is on the switch to a new cob, for the change is seldom an easy one, as Dale and others made plain to us. Much of what makes it so hard is that a new self must do the rolling. When I speak of "sustainable farmers" I am speaking not so much about the farming practices that particular farmers follow, although that is indeed part of it. I am, in the main, speaking about those who have come to identify themselves as sustainable farmers and to feel a phenomenological comfort in doing so. Most sustainable farmers will be the first to tell you that their actual practices don't match what they believe sustainability to be, as we'll hear later. But they're committed to trying, within the particular circumstances in which they find themselves.[3]

And that commitment is based in part on a commitment to a new sense of self and its fabric, its weave of social relations and social interests.

But there is something in the sense of trust in a knowledge cultivation that encourages its adherents to feel its origins in something beyond the social, particularly when it is the social that seems to have ripped up a previous one. Perhaps this is in part why, when sustainable farmers spoke to us about how that commitment to a new self came about, they, like Dale, often did so in theological terms, as a sudden conversion through a personal encounter with a higher authority. One Iowa sustainable farmer even says that he heard a voice from the sky one day while up on a ladder painting his barn. Dick Thompson, the famous Iowa sustainable farmer, had a similarly remarkable experience. He put it this way during one of the many field days he and Sharon Thompson have hosted at their farm over the years, all seventy-five of us on the edge of our seats in the family's machine shed.

"Now, this is something that is hard to be explained. I've never done this before at a field day."

Dick looked out over the crowd. I felt that he was looking right at me. Most everyone else in the shed probably felt he was looking right at them too.

"In January of 1968, while chopping stalks in field number six, going north, I was—I'd had it. All the work. The pigs were sick. My cattle were sick. I hollered 'help.' That's about the only way I know how to explain this. But some things started to happen. And a lot of things that happened seem to happen early in the morning. That thoughts come into my mind that I know that are not mine. So I want to share this. The creator wants to put a receiver, a still small voice, way down deep inside each one of us, for communication. It's our choice. It's not forced on us. If you want it, you can have it. If man can send pictures through the air to our TV, in our houses, and if man can send voices through the air from one cell phone to another, it shouldn't be too hard to understand how the maker of mankind can do the same, and much more. A personal communication. No hackers. No one to hone in. There's much that's new about the Internet and the world wide web, and all this information. That's good. It can be confusing. But I want you to remember that the Internet is a tool, not the toolbox. It's not the source. Don't let W W W be a substitute for the still, small voice."

It takes a special fortitude to talk about such a thing in public, even for a seasoned public speaker like Dick Thompson. The Thompsons are a very public farm family, of course. They figure that more than eight thousand people have visited their farm to learn about the changes that they have made since that day when Dick hollered for help. When they arrive, the Thompsons offer them a two-hundred-page report about their farm—*Alternatives in Agriculture,* they call it—which they update every year. Although Dick had never spoken about it before at a field day, the story of their break, their rupture from what most others farmers did and do, has long been in the report, right in the first chapter:

> The real change started taking place in 1967, when we began learning about the Holy Spirit. This is when Dick realized that he was caught up with things, building a kingdom with sheds, silos, cement floors and more land. Enough was never enough. We were to the place where we were looking for something better. The livestock were always sick with one disease or another. Sickness was the rule and health was the exception. A word came to us in a supernatural way, through the gifts of the Holy Spirit, the word being that God was going to teach us how to farm.[4]

Whatever theological implication one takes from such stories—I, for my part, am not a believer in such a directed notion of spirituality—one thing is clear: that the change to identification with sustainable agriculture is commonly experienced as an intense, rapid, holistic crossing over. So much is at stake. A self. A farm. A way of knowing and doing them both. One wing doesn't allow you to take off suddenly and lead a different life. Indeed, a single wing would be a considerable nuisance, unbalanced and dragging in the dirt. I think what Dick and Dale are saying is that you need them both at once, the whole set, if you are to lift off from a previous mode of being, a previous phenomenology of farming, a previous structure of practice and practice of structure.

Dick and Dale are saying something else too. They are describing a special kind of faith in sustainable agriculture's cultivation of knowledge. When they identify sustainable agriculture with a still, small voice that is external to the human, a direct voice that can't be honed in on and that is safe from hackers, they are talking about a *natural* sense of the true cob. By "natural" I don't just mean ecological relations unpolluted by pesticides, nitrogen fertilizer, and humanity's attempts to manipulate the earth. I mean their sense of an ecology of knowledge unpolluted by humanity's attempts to manipulate the truth. I mean their faith that they have encountered a source of conscience—what in other work I call the *natural conscience*—that they regard as apart from the social and its dark interests and power plays.[5] I mean their sense that they have discovered a realm of self that, finally, they can trust.

But such a discovery is not a matter for prediction by the crystal ball of social theory. Over the past ten years or so, a number of surveys on sustainable farmers have been conducted, looking for the statistical factors that might help predict why one person turns to sustainable agriculture and another does not. They haven't found much. Along with my colleagues Gordon Bultena and Eric Hoiberg, I helped conduct one such study myself in 1995, a telephone and mail survey of 705 Iowa farmers.[6] We didn't find much either. Sustainable farmers are unremarkable in their age, their educational attainment, their political affiliations, their ethnicity, and their household sizes and structures.[7] Their farms are noticeably smaller, by a couple of hundred acres, on average, in Iowa.[8] But given that smaller farms is one of the principles of sustainable agriculture, this is hardly surprising. In sum, the social landscape of Iowa agriculture is more remarkable for its evenness of pitch and aspect than for its rugged contours of difference.

Our Iowa survey did find that sustainable farmers are, taken as a whole, more religious than conventional ones. But we had no information on how religious they were before becoming sustainable farmers. As with Dick and Dale, the switch to sustainability seems to be something that many sustainable farmers understand through religious language, at least in part. It thus stands to reason that sustainable farmers would be, on the whole, more religious. But that doesn't necessarily tell us why they became sustainable farmers. It may only be telling us *that* they became sustainable farmers.

There is also some evidence to suggest that farmers are usually in their younger years, typically under fifty, when they make a commitment to sustainable agriculture.[9] This, too, stands to reason. It may not have much to do with the inquisitiveness of the young, though, for most young farmers are evidently content to direct their inquiring minds toward the subtleties and challenges of the Big Tractor, Big Chemical way.[10] It is probably an economic matter. Older farmers are more likely to be financially secure and thus less susceptible to the economic stress that so many of the sustainable farmers we spoke with emphasized.

Economic stress is, of course, part of the standard crystallography of theories of social change, and I have tried to emphasize its importance throughout this chapter thus far, albeit arguing for a phenomenological understanding of economics. Yet even here there is need for much analytic caution. For many of the sustainable farmers we spoke with, the mid-1980s farm crisis seems to have been the economic straw that broke the phenomenological back of their previous style of farming. It is probably no accident that PFI itself was founded in 1985, at the height of the 1980s farm crisis. For Dick Thompson, however, the crisis came two decades earlier. For Dale the crisis came a decade later. There is, then, a lot of what statisticians call "scatter" here.

Which points to an important observation which is all too often overlooked: Farming has been in perpetual crisis since the 1930s at least. And even in good times, the grace of good economic fortune is not universally shared. In bad times, most farmers face substantial economic stress—that, after all, is how we define bad times. In other words, there can hardly be any farmers, sustainable or conventional in their allegiances, who have not felt the shake of the economy's fault lines.

Nonetheless, most farmers have not become sustainable farmers. We must take this as a basic similarity of sustainable farmers and those who cultivate a more conventional row: The uncertain landscape of farming is something with which they all must contend. Altogether, this doesn't give the sociological crystal ball reader much to go on.

Which probably means that the decision to identity with sustainable ways of doing and being is not a crystal ball kind of thing. I found myself talking about this with Elwin and Joy one afternoon in the cluttered farm office out in the machine shed where, on an earlier visit, Elwin had spoken to me about how industrialized agriculture leads to a loss of rural community.[11] Elwin and Joy have been working for the past fifteen years to switch to a different approach to their agriculture. They've built hoop houses for their hogs. They do ridge tilling. They've instituted a longer crop rotation. They have a small herd of sheep. Joy has also started up a small "community-supported agriculture" venture—what is usually called CSA for short—in which she raises vegetables for a number of local families, delivering them a weekly load of what's in season throughout the summer. Theirs is not an organic farm. They regularly use herbicides and other farm chemicals. But they're able to do so at a far lower rate than "the neighbors," to use the phrase farmers often use to describe the broader farming community.

Elwin and Joy described the changes for me over instant coffee mixed with packages of powdered dairy creamer that they keep out in their machine shed, along with an electric hot water urn big enough for a church supper.

"So, this is a farm that's undergone some big changes," I said, when they finished the verbal tour of the farm operation and its history. "What brought about those changes? What led you to consider something else?"

Elwin and Joy exchanged glances. This topic seemed to be mainly Elwin territory, they appeared to decide. At any rate, he took the lead in addressing it. "Well, it's no big secret that the bank panic and the land price crash in the mid-eighties nailed us hard. That was sort of a—I don't know what you'd call it—the knock on the head that maybe the system we were using wasn't the best. Because we saw up close and personal how fragile that system was. So definitely the bank panic, which was also a result of the land price crash, is what stimulated us to look for something different."

I nodded, both to them and to myself. An economic "knock on the head"—that sounded to me like phenomenological rupture of the structure of practice and the practice of structure. In fact, in was during the course of this interview that I began to suspect this might be a common pattern among sustainable farmers.

"Yeah," said Elwin, acknowledging my nod. "So it'd been eighty—"

"Early eighties. Eighty-two, eighty-three," Joy interjected.

Elwin turned to her. "The Practical Farmers of Iowa was founded in eighty-four or eighty-five or something," he said, trying to sort it out.

"I think in eighty-five," I suggested.

"So the timing," Elwin continued. "We were looking for something different about the same time PFI was founded."

"You were one of the ones that went up to one of the satellite meetings in Powell Center," Joy remembered.

"I think that was the year after PFI was formed," Elwin said. "Dick Thompson and some other people held three meetings across the state in the summertime, kind of exposing people who had a philosophy defined by industry to something different."[12]

He looked over at Joy. She nodded.

"If that's your question on what did it," Elwin concluded, "it was the land price drop in the mid-eighties."

But that didn't fully answer my question, I realized as I listened to Elwin and Joy.

"As you're saying this, I'm thinking to myself, but there were certainly other farmers who got caught just as bad in the land drop and who didn't go sustainable. Makes me think that there was something else going on."

"Yeah. You're right. There certainly were others who stayed conventional," said Elwin.

"And are they still in farming?"

"Oh yeah. And they're still conventional."

"So what was different about this family, this household? Why did you, I mean, maybe you don't know the answer to this question." This was really an impossible thing to ask them. If someone were to ask me something similar—like "Why did you become a sociologist when so many other people have not?"—I haven't a clue what I'd say in reply. But after a moment's thought, Elwin and Joy came up with a pretty sharp response.

"You mean why did we do it when others didn't? Why did that light bulb come on? I'd have to think about that one. I don't know. You've given me something to think about. Joy, what do you think?"

"I don't know, really. I guess it gets back to how you were raised."

Elwin nodded. "Patterns that were set in your mind a long time ago. And then one day it makes a difference."

In other words, as in a good conversation, things happen to us in social life that we do not expect. Part of what makes a conversation a conversation and not a script read out loud is that we find ourselves hearing things and saying things that we did not—and could not—anticipate. There is a *dialogic unpredictability* to the course of human lives. Indeed, that is what gives life to human lives.[13]

But dialogic unpredictability is not randomness. We have some choice in the conversations we enter into, and thus some reasonable expectation of how our conversations may turn out. We choose these conversations in part based on those expectations, although we are sometimes surprised by the talk that actually unfolds, occasionally very surprised. And we also lack some choice in the conversations we enter into, given what we know and don't know, and that lack of choice may also give us some reasonable expectation of how our conversations may turn out. Moreover, the history of the conversations we choose, and the conversations we don't choose, sets patterns in our minds, as Elwin noted. Maybe "patterns" is too rigid a word for it, though. We are who we are because of the *living sediment* of our life's conversations, a sediment that settles out of the tumbling currents of sociality and the riffles of our own words and deeds. The living sediment: It's an accumulation that never stops stirring around, as long as we are alive.

Meanwhile, others are making their own choices about the conversations they wish to enter into, as well as being subjected to their own lack of choice (which, indeed, is often the result of the choices of others). Fluid dynamics is a complex enough physics on its own. But this is a fluid dynamics of wills, billions of them, swimming, jostling, playing, thrashing, trying to find some traction on the stream bed, and discovering now and then that their struggles with the current have brought them within range of new conversational possibilities that seem suddenly relevant, handholds and footholds for a body in need of air and perspective. And so a few of them decide to go to a meeting in Powell Center to hear Dick Thompson speak about a whole new stream on the other side of the bank.

Metaphors can get in the way of understanding, as a writer tries to jam the world into something it is not. So let me put the matter again, more plainly. PFI farmers are far more pronounced in their similarities with other Iowa farmers than in their differences. What, then, accounts for why economic stress leads to phenomenological rupture for some farmers and not others? Probably not the degree of stress. The stories of economic hardship we heard from PFI farmers did not seem remarkably severe in comparison to other Iowa farmers we spoke with. Nor do the standard demographic factors that sociologists often consider have much to do with the turn to a new phenomenology. It's, well, luck, but luck of a special kind. Let's call it *dialogic providence*—a situational luck that, at least momentarily, perks your ears for conversations from more distant knowledge cultivations, cultivations whose words just happen at that very moment to come drifting your way, or perhaps do not.

Economic stress was not the only source of phenomenological rupture that sustainable farmers discussed with us. Many also stressed the importance of health issues.

Dwayne is one of them. Dwayne is one of the two sustainable farmers I had talked with about farmers' identity at that field day at the farm of Dave Lubben, the former PFI president—the day I called soybeans "button weed." At least my gaffe that day made me stick out in Dwayne's mind when I later called him for an interview. He remembered me, all right.[14] Sure, an interview would be fine.

Dwayne and his wife Jody farm about two hundred acres in the hillier section of Powell County—corn, soybeans, and a small "cow-calf" beef herd. (A "cow-calf" herd means raising beef animals from birth, as opposed to buying animals to "feed out," as in a feed lot.) It's not enough to support their family of seven, so she has a job running a cash register at the Walmart in the next county and he drives a school bus every morning during the school year. They used to have the thousand-acres dream, and he was carpet bombing the fields with all the standard farm chemicals, trying to get there. But they gave all that up about ten years ago. Now they "band" their herbicides along the rows and cultivate between the rows, reducing their herbicide and insecticide costs to less than $5 an acre on most of their fields, well below the standard $40 or so that most Iowa farmers spend.[15] They also rotate grass with their grain crops, using rotational grazing techniques for their beef herd when they do—rotational farming in two senses. Dwayne and Jody even farmed organically for several years, but gave it up because "weed pressure was starting to build to the point where I couldn't control it anymore," as Dwayne explained. Their goals are more modest now: keeping their costs low, being good stewards of the land and the water, and farming in such a way that they don't feel pressured to try to take over the neighbor's land. Even though they're not organic anymore, Dwayne and Jody made it very plain that they have no intention of going back to the way they used to farm.

When I arrived in the late afternoon, Jody was still at work. Dwayne had started to cook supper for the kids, and throughout the interview they periodically tumbled out of the TV room and into the kitchen, where we sat at their battered Formica kitchen table, to inspect the unusual visitor and to ask the local judge to resolve their various disputes—and at one point to get a bloody nose mended. The compressor on their old refrigerator fairly roared, and the cuckoo clock on the kitchen wall kept up a steady tick and quarter-hourly song. It was the kind of interview that gives tape transcribers nightmares.

The cuckoo clock was not the only avian presence. The kids let the family cockatiel out of his cage, and he took up a perch on Dwayne's shoulder. Dwayne let him stay.

"I just wanted to let you know," I said, "I never interviewed somebody with a bird on their shoulder before."

"You're making history, Spike," Dwayne replied, stroking him under his beak.

I laughed. I'd never met a cockateil named "Spike" before, either.

Dwayne proceeded to describe the remarkable history of how they came to leave the monologic, Big Chemical way behind. Back in the eighties, a mysterious illness wiped out their cow-calf herd three times in three years. First the cows would start losing weight, then they would stop producing milk for the calves, and then they would start dying. The local vets couldn't figure it out. The county extension agents couldn't figure it out. The specialists at the National Animal Disease Laboratory couldn't figure it out. After the third time, the banker wouldn't give them the loans for a fourth herd. So they concentrated all their efforts on the small farrow-to-finish hog operation they had at the time. And then the same condition hit the sows. They couldn't keep their weight, and their litter size went down to one or two piglets. Needless to say, Dwayne and Jody's finances were by now wildly out of balance, and they were about to lose the whole farm.

It didn't seem like a disease, though, if the same condition affected two livestock species. Dwayne wracked his brain for what could have changed in the farm operation. Finally he hit on it, he explained to me. One section of their farm lies at the bottom of the local drainage system, and one field in particular. It's pretty wet ground most years, and Dwayne finally gave up trying to raise grain on it, converting it to hay ground for his growing livestock operation. Previously, the yield of that field went to the local grain elevator. Now it was going to their animals in the form of hay, and immediately they fell sick. Dwayne concluded that that field, lying as it did, was the settling point for all the pesticide runoff from their farm and their neighbors' farms. He was poisoning their cows and their pigs every time he fed them hay from that field. (One part of this story I did not fully understand is that it is rather unusual to feed much hay to pigs; I didn't think to ask Dwayne about this at the time.)

This is an amazing story, and there is probably no way of verifying it, one way

Sign at the Jedlicka farm, PFI *members, with "hoop houses" for hogs in the background, near Solon, Iowa, 1999.*

or the other. It was years ago, and Dwayne has put that low-lying field in the USDA's wetland reserve program now. He isn't feeding the hay from it to their livestock, and the cows are not dying on him anymore, for whatever reason. As far as the topic of this book is concerned, what matters is the change this experience wrought in his and Jody's outlook on farming.

"When you realize that the cows were—that you lost a prize herd because of poison, it really makes you think," Dwayne explained. "There's gotta be a better way."

A shock like that, a deep rent in one's fabric of trust, indeed must make one think. In Dwayne's case, it led him to question the monologue of Big Ag he had long been an audience for. Dwayne started perking up his ears for those more distant knowledge cultivations, and he soon found himself an active participant in them.

Health issues are notoriously hard to verify with the meters and computer printouts of "objective" science. In addition to the continued inaccessibility of much of the body to the probes of high-tech medicine, our feelings of health are in themselves real conditions that nevertheless no meter or computer program can detect. I write here of broader issues of one's confidence in the world and one's place in it. These are matters of the mind and not the body, goes a familiar response. But the mind is not just connected to the body; it is part of the body, and we know this at a deep level of our daily experience. These deep levels aren't easy to talk about, though, and the language of health is often one of the means we use to give voice to our mental certainties and uncertainties. Which brings us back to the economic. While several farmers stressed the importance of health issues in their decision to switch to identifying with sustainable agriculture, it was evident from their explanations that these health crises were simultaneous with economic crises.

To repeat Dick Thompson's holler for help, "In January of 1968, while chopping stalks in field number six, going north, I was—I'd had it. All the work. The pigs were sick. My cattle were sick. I hollered 'help.'" *All the work*. Dick is describing a broader crisis here than the merely medical.

Dwayne said as much, too. "Second of all, by that time we were so strapped for cash, I had to come up with some other ways. And I started experimenting with cutting back on herbicides."

Of course, if your cow herd dies on you three times in a row, it's going to cause pretty severe financial problems for your farm. That's part of my point. If the bank had never felt the need to call Dwayne's cow-calf operation to a halt, his cattle's sickness most probably would not have troubled him so much. Indeed, livestock on most Iowa farms often die before farmers can sell them. Many farmers even buy "livestock mortality insurance" (despite the contradictory name of such policies); and there is much interest among farmers in new regulations that allow inexpensive on-farm composting of "dead stock"—a surprising term when one first encounters it, but a common bit of agricultural talk.[16] "If you have livestock, you're going to have dead stock," goes one Iowa country aphorism.[17] As long as the banker keeps handing out the loans.

And let us not forget that most Iowa farms are commonly in economic ill health as well. It may be—although I think this would be a hard matter to determine with certainty—that a sense of economic health pervades one's sense of medical health more broadly, even if the health problems of one's livestock and one's own self and family are not unusual in the eyes of an external observer. It may also be that economic health influences that deepest sense of health, one's sense of spiritual health.

But even if a sense of economic health and medical health are sometimes linked, and even if economic and medical ill health are common on Iowa farms, the fact remains that these stresses generally do not result in farmers converting to sustainable agriculture. Social science still has no warrant here, and perhaps not anywhere, for the crystal ball of prediction, at least at the level of the individual and his or her conversations of life. Discovering and tuning into the conversations and cultivations of sustainable agriculture remains a matter of dialogic providence. But providence is not always gentle.

Many PFI farmers—at least a third, I would estimate warily—told a different kind of story.[18] These are the farmers who have been trying to farm sustainably ever since they began farming their own farms. For these farmers, the change in phenomenological commitments happened before they began their own farms. PFI farmers described a wide variety of paths to these changes before their farming days. But especially common was an extended period in their twenties or thirties spent living overseas, usually in a developing country, perhaps as a teacher, a Peace Corps worker, or a missionary in a religious-based development organization. Mexico. Costa Rica. Panama. Columbia. Bolivia. Haiti. Malaysia. The Philippines. Japan. Uganda. Zaire. Swaziland. These are all places PFI farmers have had extensive experiences in their pre-farming years.

Take Jon. He's a former antiwar and antinuclear activist, now in his late fifties. He was raised in a city and has a master's degree in political science from the University of California, Berkeley. But today, along with his wife, Heather, he is raising organic beef on a 320-acre farm in the western Powell County hills. They've been farming for about eight years now. It's a gorgeous place: rolling pastures, a backdrop of woods, and a lovely stream. But the house is a wreck. Shingles missing. Porch sagging. Paint peeling. It was even worse when they arrived. The toilet fell right through the floor into the basement one night their first year there. Jon and Heather bought the farm because it was cheap and beautiful, not because it didn't need work. Neither of them had ever farmed before, and they didn't know anyone in the area. They saw the place advertised in a national farm real estate listing, and it fit their budget, so they bought it and moved in. After a tough climb up a steep learning curve, they're finally feeling settled and in reasonable control of what they're doing. Jon's even considering giving up his job in town and going full-time on the farm, now that they've hooked up with a marketing cooperative for their organic beef.

But the real upheaval in Jon's life happened long before they arrived in Powell County, when he joined the Peace Corps at age twenty-one, three months after finishing college, in the late 1960s.

"I lived out in the middle of nowhere in a mud hut," Jon said of his time in Honduras, teaching math and science in a village school. "I just realized how wonderful it was to just live out in nature. I didn't need that much. I really don't need all these things, you know. And the idea also occurred to me that I don't need my culture. I first was in culture shock, needing my culture very much, and living in this alien culture. Then it started to creep into my soul. It was sort of like osmosis, pulling this culture into you, this foreign culture, this different culture . . ."

"Sounds like you were learning as much from them as they were learning from you," I said, sipping the spiced Indian tea that Heather had served me.

"Oh, that's true. Absolutely. More so. Once I began seeing my culture as an outsider, it was a very—almost a very, how would you say, a scary, scary experience. Because when I went down there I was twenty-one, right out of college, beginning to look at my culture from the outside. I was there to help those people, and to get to know those people, and learning to speak their language. And it was so radically different. I mean, these people had nothing. A mud house and a few chickens and no implements but a machete. . . . I began to see a lot of things that, the experience was—I mean, it was a watershed."

"So you didn't have a draft issue?" I was thinking that somehow his Peace Corps work kept him out of the draft, not knowing very much about these matters, being too young myself to have experienced them.

"Oh yeah. Heavy duty. And, another thing, the experience down there made me totally against the war. These people were the same people like the Vietnamese. I mean, living in the same kind of situation. And I thought, how can I go and kill these people? If they wanted to be communist, I didn't care. It was better than what they had. It seemed like a non-issue to me."

As with so many others at the time, the Vietnam War was a terrific phenomenological wrench for Jon. But it was a second wrench, coming as it did on top of his Peace Corps experience. It made him doubt the value of his Peace Corps work. Even more, it made him doubt just about everything he once knew and thought he could count on.

"I began to realize that the U.S. was not serious about what we were doing. I mean, even as Peace Corps volunteers, they were not serious about helping us [do our work]. It was just window dressing. They wanted to get these young kids and send them down there to look good. If you were brought up in the fifties, which I was, America is great. We're number one in the whole dang world. And then all of a sudden you realize that, hey, they've been lying to me. I mean, like, big time. There's something here that doesn't add up. And then you try to talk about this, and people are looking at you like you're crazy. Or they call you a communist. And you're going, it was just emotionally—I mean, it was very hard for me. Very, very hard for me. What it opened up mentally was, 'Okay, what else isn't true?' "

This experience of recognizing monologue—"they've been lying to me"—built a deep sense of mistrust in Jon that left him reeling for many years.

"I actually got to the point where I kind of exploded all my priorities, and I kind of had a nervous breakdown. It was literally like that. I mean, I was unable to see. You don't see that house," he said, pointing to a photo on the kitchen wall. "You didn't have categories to associate information together. I mean, I actually got to that point. . . . I was dying, I mean literally. I was really in bad shape."

He read widely, compulsively, searching the Western canon for a truth he could trust. For a long time, he couldn't find that firmer ground, he said. And then what finally pulled him out of this phenomenological swamp was his discovery of a natural conscience—a realm that he felt to be beyond the prevarications and power moves of social life.[19]

"I was doing this study, and I came to the conclusion that I had to abandon the intellectual side. I had to abandon looking for truth that other people search for—Kant or Hegel or Locke or whatever. I had to say, 'Okay, in my life, what was real for me? What was true?' And so, the one truth that I could come to was the natural world. 'Okay, what is true? What is true? And what is true that I know is real?' And that's the natural world. And it's God, and how to experience God. I mean directly—not God through the Bible and interpretation and somebody telling me about it. . . . It's like I had to go out and reexperience the world. Reexperience my cultural relations. Kind of go through them from a whole different sense."

He paused finally. The words had been flooding out of Jon. He really needed to say this stuff. I was transfixed. Heather had heard most of it before, I'm pretty sure. But she seemed to feel there was important catharsis in the re-telling, and said nothing, except for here when she helped Jon along.

"It was affecting you physically too," she reminded him. "Your back, you could hardly move. You had severe back problems."

"I had such horrible back problems," Jon continued with a nod. "I couldn't get out of bed. I was just like an invalid. I was holding myself back, just literally holding myself back. It was like a metaphor, illness as metaphor. It was weird, but it was exactly like that. But I had to find this truth. And as I worked with this truth, and worked with it, I had to work with my own physical body. Because, I mean, I was in a lot of physical pain and I didn't know how I was going to make it economically. Because I could hardly work. Could hardly walk."

The resonance of health problems, economic problems, truth problems, and spiritual problems, I thought to myself, hearing here a further resonance with the conversations I'd had with other sustainable farmers. "Illness as metaphor," as Jon shrewdly put it. And like many other sustainable farmers, Jon sought a refuge in a truth he felt to be beyond the lies, the posturing, the window dressing of social life—in a knowledge cultivation that he felt to be rooted in nature's own secure conscience.

"When you're working with nature you're working with reality—and not all the bullshit," Jon continued. "I was really scared. It was like my life was at stake. It was like either die, or do this. It's the only way I would have done it. I had to radically change my life. . . . I was going to have to get into agriculture. Something where I was going with the natural world. I needed to live with living things."

Heather's story followed another course. For her, there was no dramatic breach with the past, no rift in the landscape of her personal history. Heather came to sustainable agriculture through a progressive unfolding and refolding of the layers of ideology. Hers was a peopled story, a story of personal relations of knowledge and trust, of those who fortuitously helped her along in the unfolding and refolding, from her own mother to some "back-to-the-landers" to Jon. We got a chance to talk about it when Jon headed outside to check on the cattle, Jessica, their young daughter just home from school, in tow.

"Jon had this big break, almost kind of a rupture from his upbringing," I began, using the word "rupture" that had been so much on my mind while listening to Jon. "How about yourself? Did you go through something that bad?"

"No," she replied in her quiet way. "I didn't come at it at all from the same way that he did. . . . I think for me—I was trying to think where it all started."

Heather paused to collect her thoughts. "I lived in this rural community 150 miles north of San Francisco, and in the early seventies they started having an influx of the hippy type coming from the Bay Area. The back-to-the-landers. And my mom—my dad was and still is a lawyer—and my mom at that point, well she was always into astrology. A health food store opened, and she started getting medicine from these people. She started getting involved with them, and they started coming over to our house. We had two women who had moved—probably originally from the Midwest somewhere—that lived in San Francisco and then moved up and lived in this old school bus on this land. Back-to-the-landers. They lived with us one winter because they didn't have any access to water, I guess."

"And your parents just got to know them and thought it was—?"

"Yeah. My mom got to know them through the health food store. They worked there. My dad was a little leery at first, but I think when he got to know the people, he liked them. And then he started feeling more comfortable. But initially it was my mother meeting these people and starting to bring them [home]."

The back door opened and their daughter Jessica stuck her head through the doorway between the kitchen and the mudroom, carefully keeping her boots in the mudroom (well-trained child). She pointed furtively to an old aluminum pot hanging on the kitchen wall, evidently needed for something having to do with the farm work—perhaps for mixing some kind of feed supplement for a calf, I guessed. Heather got up and reached it for her. "Can you bring that back?" she called to the fast-retreating figure, disappearing in the slam of the screen door. I gathered it wasn't the first time this pot had made its way out to the barn.

"And so I think that my first connection was kind of through that," Heather continued, settling back into a kitchen chair. "And also through the food aspect. Because these friends were into health foods, you know, and whole foods. And so my mom started getting that. I was probably interested in foods somewhat before that anyway, but I started getting interested in the health aspect of foods."

"So what year is this, more or less?"

"In the mid-seventies."

Heather is quite a bit younger than Jon and, like me, missed the main fireworks of the antiwar movement. But there's a definite sixties, hippy feel to her story. Perhaps her mother had gotten caught up in the turmoil of those times somehow, and this was the source of the leeriness of her lawyer father. Perhaps her mom had had something of the same kind of rupture as Jon's, and then decided to raise Heather along these somewhat countercultural lines. I didn't think to ask at the time. I don't know how much she would have told me about her parents' relationship and history even if I had asked.

But Heather did tell me about how from these experiences in her late teens she decided to get a degree in restaurant management and become a natural foods chef. She worked for a number of years in a West Coast natural foods restaurant chain, until it was bought out by a major food corporation.

"That tea was used in the restaurant," she said, pointing at the pot of spiced Indian tea I had been enjoying so much. She met Jon through the restaurant. He used to come in Saturday mornings for breakfast, she said, "and somehow we started talking . . . about how he had been looking for land, and somehow we connected with that. My dream at that time was I wanted to have my own restaurant and grow my own food, like be in a rural setting, raising my own food and making everything myself. And just having a real small restaurant with a seasonal menu. So that is what I wanted."

Heather doesn't have the restaurant yet. Maybe someday. In the meantime, she said, "We're both committed and we're both really glad to be here. It's been a really difficult life for us, but neither of us would change that."

Stella's story traced yet another path. She and her husband Lyndon now operate an organic fruit and vegetable farm. They run a CSA, and during the summer they make the rounds through a different farmers' market almost every day of the week. They're active members of PFI. But back in 1985 they were struggling to hold on to even an acre of what had been a conventional, midsized grain, tractor, and chemical farm. The drop in land values associated with the mid-eighties farm crisis hit them as hard as it hit anyone. By the end of it, Lyndon was working in town and Stella was looking for something to do.

"So we started growing vegetables," Stella explained to me matter-of-factly from across the table at the conference dinner we were both enjoying one winter at a PFI meeting. PFI tries to serve local food raised on member farms at its events, and the meals are often outstanding. Some of Stella's produce was on the menu. Given the incredible energy she puts into her farming now, it's hard to believe that she was ever looking for something to do.

"Did you do it without chemicals right from the start?" I asked, forking a tender red potato, possibly from Stella and Lyndon's farm.

"Yes, but not because we were trying to farm organically."

I looked quizzical.

"It was just the way my mother had always done it," Stella went on. "She had always kept a vegetable garden in the backyard when I was growing up. So it was just the way that I knew how. Later on we found out it was called organic!"

There's an old theory of the spread of technology and new ideas called the "adoption-diffusion" model. The theory was originally developed to explain, and to promote, the spread of hybrid corn in the 1950s.[20] The basic idea was that one could predict which farmers would be "innovators" and "early adopters" right through to "late adopters" and even "laggards" (one of the more loaded terms of social science) based on "adopter characteristics" such as education, income, social status, farm size, and the like.[21] The theory is now questioned by many observers of the development of sustainable agriculture, however, in part because of its strikingly arrogant presumption that cutting-edge ideas start at the top, generally at the university, and diffuse downward based on adopter charac-

Adam and Dennis Hansen feed sheep at their farm near Audubon, Iowa, 1999.

teristics.[22] Stories like Stella's turn this theory completely on its head. Here the ideas of a laggard on the bottom—Stella, who, at least when it came to vegetables, had never learned about chemical farming—turn out to be among the cutting-edge ideas of the present day.

Stella and Lyndon definitely went through a phenomenological rupture, precipitated by the rural economic storm of the mid-eighties. Yet they turned to an older conversation, an older cultivation, for a new foundation of life practice and life structure. They didn't go to hear Dick Thompson speak at that Powell Center meeting in 1985. Perhaps they felt no need to. Perhaps word of it simply didn't drift their way.

In any event, this older conversation connected easily with the new language of sustainable agriculture. In time, Stella and Lyndon have come to identify with that new language, that new cultivation. They have found new words—such as "organic"—for practices that were old to them, as well as new words for practices that were indeed new to them.

Most notable among those new words and new practices, Stella told me, was

CSA: community-supported agriculture. Although Stella and Lyndon continue with the farmers' markets that have been the mainstay of their income since they switched to vegetable farming, their CSA customers are increasingly the focus. As Stella put it later in our conversation, her eyes shining, "CSA has changed my life."

They're rolling a new cob now.

A different sustainable farmer, a different story. A different conventional farmer, a different story. Herein lies the humanity of every farmer. There are common features in many of their life stories, in the history of their life conversations. I have tried in this chapter to point to the most prominent ones: economic stress, feelings of ill health, doubts about truth, and spiritual doubts, leading to phenomenological rupture and the recognition of social interests and tuning in to a new cultivation of knowledge about the economy of the self—about the structures and practices of the self. But these are not matters for equations, computer programs, and robotic certainties. In this there should be no sociological lament. Quite the contrary; it is cause for celebration. That the outcomes of a life's conversations cannot be predicted is what gives a person her or his unalienable aliveness.

Let me say it again: Most farmers have never switched to sustainable agriculture, or chosen it to begin with, despite the near certainty that points of severe phenomenological stress have been features in their lives as well. It is not an easy matter to recognize monologue. It is not an easy matter to reject monologue. Some do; some don't. Just as the same seed in different ground will produce plants of different form and fruit, so common social factors will lead to different outcomes in different lives. Indeed, as every gardener knows, the same seed will produce plants of different form and fruit even in what appears to be the same rich, silty loam. Such is the dialogic providence of the soils of social life.

7. Farming with Practice

"So why is it saying—" I pondered, looking from the passing road sign to the map on my lap.

South 245. Yes, why is it saying that?" Rick Exner tipped his blue Practical Farmers of Iowa cap up his bald head, sharing my wonder at the way back roads often seem to shift around of their own accord.

"Why does it say that? So we must have gone right past County Road W, right? I think we missed W." We were losing our grip a bit.

With some reason. Rick and I were running late for the PFI field day at the Powell County farm of a PFI "cooperator"—a farmer who conducts sustainable agriculture research in cooperation with other PFI farmers running related trials, often in conjunction with university scientists and often on their own. "On-farm research" is what they call it. Rick had organized the field day as part of his job as on-farm research coordinator for PFI. It's a job he's held since 1989, when he became PFI's first (and, for many years, only) employee. Rick had the PFI sign, the literature, and the group's portable sound system in the back of the car. So we really needed to be there pronto.

"Well, there was a sign a ways back," suggested Rick, glancing from the road toward the rumpled map I was poring over and simultaneously trying to keep track of his driving.

Rick has steady hands, so I wasn't worried. And he's an experienced map reader. He has been one of the main steerers and map readers for PFI since the group's inception, and he knows the Iowa rural landscape as well as anyone. I'd been trying for months to sit him down for an interview about the history of the organization. He suggested that I ride down with him to the series of PFI field days being held that day and the next in Powell County. "Bring along your tape recorder. I'll be sort of like a captive audience in reverse," he joked. It seemed an excellent plan to me.

One of the things I was eager to ask Rick about was the origin of PFI and its curious name—*Practical* Farmers of Iowa. I was struck by it the first time I heard it, and a number of people I've told about the group expressed similar curiosity about its quirky down-homeness. I'd heard Dick Thompson say once that Rick had come up with the name, back in 1985, when Rick was finishing up the

Discussion in the machine shed at Maria and Ron Rosmann's farm near Harlan, Iowa, during a PFI *field day, 1998.*

research for his master's degree in agronomy at Iowa State. Rick's research was based on some field trials of inter-seeding corn with a legume crop—clover, in this case—as a form of nonchemical "green manure," in which the legume's nitrogen-fixing capability fertilizes the soil for the nitrogen-hungry corn crop. As it happens, the trials were conducted on Dick Thompson's farm. This was not "on-farm" research in which farmers were active participants in the trials. It was simply research on a farm, more engaged with the real world than the usual controlled studies on university land, where most agronomic research takes place, and something the farmer, Dick, was interested in. It was academic-led research nonetheless, but it turned out to be a way for Rick and Dick to get to know each other better—to discuss their mutual sense that there were other ways to farm than the industrial way, and that there were a lot of farmers who already had some pretty good ideas about how to do so, Dick among them. He'd already been making some dramatic changes on his farm, such as ridge tilling without herbicides, and he had been conducting some rudimentary experimental trials on whether it was really working. After all, the university researchers at the time were doing next to no work on whether there might be other ways—aside from limited studies by Rick and a few others. Maybe there was a need for a group that shared the knowledge of farmers and these few others about alternatives to the Earl Butz get-big-or-get-out view.

Indeed, that was exactly what Larry Kallem, the executive director of the Iowa Institute of Cooperatives, suggested to Dick. At a series of lectures at Iowa State on biological approaches to farming, Dick and Rick raised the ideas with the farmers in the audience, and hands of people who wanted to join such a group went up everywhere. Larry helped them get it going, and has remained an advisor to the group ever since.

"Do you remember the first meeting you went to?" I asked Rick, once we had sorted out what had become of County Road W.

"I remember a number of meetings at Dick and Sharon's kitchen or living room. I remember one where we were trying to come up with a name for the organization."

"Okay, how did that come about?"

"Well, we didn't have a name. We thought it would probably be a good idea to have a name!" We both laughed at that. "We wanted a name that was not specific to the cubby holes that people were putting alternative farming in. We thought it better not to have a name as specific as 'organic' in there, because we were looking for an organization that will be accessible to farmers of all different stripes, really. The organic farming groups have their organizations, and we needed an organization that would bring in new people, as it were. So, how do you describe what that wants to be? In terms that are positive but not exclusive? It's, as it turns out, I think no matter what adjective you pick, somebody is going to consider that they are being excluded. So, we called ourselves 'Practical Farmers of Iowa.' And then the question is, what does that make the rest of us?"

I nodded, silently rolling the phrase "impractical farmers of Iowa" around in my mind. I could see what Rick meant.

"That's okay," Rick continued. "People can have their fun with it. I think it's a good description of what the organization wanted to do and it doesn't box the organization in."

"Now, I think I had heard some place that you had come up with the name."

"I didn't come up with the name. I came up with the formulation. I mean these words were being thrown around, and Larry Kallem was articulating what we wanted the name to express. I think it was Larry who said that the word 'practical' had to be in there—that it's a good word, pragmatic, nonideological. The fact that we were grounded in what was practical for farmers to do in Iowa. We were trying to come up with something that was succinct, you know, that was accurate with those three or four words together. We tried various combinations, and then I opened my mouth and said, 'Why not Practical Farmers of Iowa'?"

Indeed, why not?

Lyle has been a PFI cooperator for more than ten years. In the beginning, he was heavily involved with a number of other PFI cooperators in testing out several variations on ridge tilling, the alternative planting and tillage technique that a lot of PFI farmers favor because of its reduced need for herbicides. (Many of them now ridge till without any herbicides at all.) It's a bit more complex than conventional tillage practices. You have to time things just right to make ridge tilling work well, so larger farmers generally find it too fussy and constraining. Bring on the power of chemistry, most of them say. And bring on the GMO varieties with built-in herbicide resistance, even if they cost more money and run the risk of promoting weed resistance through overuse.[1] But for many smaller farmers, ridge tilling's substitution of management skill for money lost to the chemical and seed dealer is very attractive. Not that Lyle's farm is all that small. He and his wife, Gail, have a four-hundred-acre Powell County spread, a bit above the average size for all Iowa farms. But there aren't that many Iowa farmers anymore who work full-time on their own land even with four hundred acres. Lyle does it by managing carefully. The experiments he's done as a cooperator have helped him make a lot of these careful decisions, he says.

But ridge tilling is old hat now for sustainable farmers. It's working well for Lyle, and he's not tinkering much with that aspect of his operation any more. Lyle is most excited these days about some experiments in manure management he's conducting with a professor of agricultural engineering at Iowa State—Lyle runs a small farrow-to-finish hoop-house hog operation—and about a dramatic new piece of equipment he's been trying out: a flame cultivator. Farmers sometimes talk about burning out weeds with chemicals. A flame cultivator literally burns them out—not by setting them on fire but by pulling a propane flamer quickly past them such that the sap boils and bursts enough capillaries in the stem to wilt and eventually kill the plant. The crop plants face the same flame, so the trick is to do your flaming after the crop has become established but before the weeds have—in other words, when the crop is tough and the weeds are weak. Strangely enough, it works. But, as with ridge tilling, you have to keep careful tabs on your fields to get the timing right.

When Rick and I arrived, half an hour late, a group of about twenty farmers, several Iowa State professors, a couple of graduate students, and an extension agent were standing in a circle under a huge oak by Lyle and Gail's house, locked

Attendees at a PFI *field day at the farm of Ron and Maria Rosmann, near Harlan, Iowa, ride out into the fields, 1998.*

in a technical discussion about Lyle's manure management trials. This was good news: They hadn't yet headed out to see the main attraction, the flame cultivator. It was almost entirely a male crowd, I noticed. The only women were an Iowa State professor and a graduate student of hers. Even Gail was nowhere to be seen.

Then Lyle loaded us up on the hayrack behind his John Deere. It was a hot day, and there was definitely a bit of wariness in the group as we left the shade of the oak. Still, the crowd happily, if sweatily, bumped along out to the field where Lyle had the flamer hitched up to a second John Deere. It was not exactly a handsome beast, crouching there in the beans—a big propane tank on wheels, with an array of hoses and pipes, one each per row, to carry the gas down to the burners in the back, hovering over the ground. It looked homemade, and it pretty much was. Lyle had put it together from a kit he bought from a farmer in Minnesota who makes them. In other words, it's not the sort of thing that shows up at the annual agricultural products show at the Iowa State Fair or at the local implement dealer's.

Lyle jumped down from the tractor cab in his PFI cap and Organic Valley T-shirt and took the portable mike from Rick.

"It's worked out pretty good," he began. "But I have to say it is hot back there," he went on, pointing to the burners.

"How fast do you run it?" yelled out someone from the hayrack.

"Pretty slow, four and a half miles an hour," which was about the speed that Lyle, a quiet and unassuming man, speaks. "And sometimes you do get a bit of a burn going, if you get some extra crop residue. Or you'll get a little smoke going when I go over one of my grassed strips."

As an organic farmer, Lyle maintains wide strips of grass around his fields to buffer his crops from drifting pesticides from the neighbors. The grass buffer strips also help the soil stay put.

Another voice. "So there's a real danger of fire, then?"

"No, no. There's not enough residue to fuel anything for long, and even the grass is just a strip. It usually burns out pretty quick. Sometimes I do get off the tractor, though, and stomp on things. But you do get a bit of smoke here and there. It keeps the neighbors talking."

"You're good at that, Lyle," chimed in the extension agent, and the crowd broke up over the well-known trials of being different in such a public endeavor as farming—the problem of living in a fish bowl, as Bert put it (Chapter 4).

"Yeah, I know," Lyle put in, only barely into the mike, after the laughter subsided.

"But let me play devil's advocate with an environmental question here," asked an older farmer. Lyle seemed to know him.

"Okay. I think I know what you're going to ask."

"That propane must use a fair bit of fossil fuel. The farmer's worried about the economics, but the consumer doesn't think about that. They're just interested in the environmental part. And aren't you using more energy?"

"I've worried about that too." Lyle's a Sierra Club member, he told me once. "But I don't think I am using more. You see, the main energy user is producing anhydrous, and I don't use any of that. So if you look at the total system, I think I'm still way down."

By anhydrous Lyle meant the anhydrous ammonia nitrogen fertilizer that most grain farmers use to jump the crop out of the ground. A fair bit of the natural gas produced in the United States goes into the process that jumps the nitrogen out of the air and into the ammonia. There was some good holistic thinking going on at Lyle's farm, I thought to myself.

"Also Lyle, how much are you pulling?" called out someone else, using farmer's lingo for the weight and drag of the rig a tractor has to manage.

"Well, that's right. I'm going slow and I don't have any iron in the ground. I'm not pulling much. So I'm using a lot less fuel."

"But that propane represents a lot of energy. I'm not sure that you're balanced out there," said one of the professors.

"But when you look at the anhydrous I'm not using, I'm still saving a lot. I'm sure I'm way down compared to the conventional guy. I'm convinced of that. But there's a lot to figure out. I'm still learning, and still making lots of mistakes with this. There are still lots of unanswered questions, like what really is the right speed, and for which conditions, and how low to put the burners for different crops."

"Did you get any instructions with the kit for it?" asked someone else.

"Yeah, it came with some stuff, and that's what I'm doing usually. Still, we really need some research done on this."

"Well, you're doing it, Lyle," said Rick, encouragingly.

"I don't know," Lyle said quietly, this time not even close to the mike. "There's a whole lot of variables there."

Ever the scientist. Or was it ever the seminar leader? Lyle took a fair bit of heat that day on his farm. He's used to heat, though. Most PFI farmers are. They like debate and difference. They don't mind thoughtful critique, even in front of others. They know that a measured bit of flame, properly directed at the weeds and out in the open where everyone can see it and discuss it, can help everyone's crops and knowledge grow. This kind of flame cultivation is what practical farming is all about.

Earlier that year, I'd had the chance to experience in a different setting the openness, the welcoming of difference and critique, the willingness to admit mistakes, and the sense of ongoing discovery and learning that are so characteristic of knowledge cultivation in PFI, at least in its best moments. A group of five PFI farm couples get together once a month through the year to talk about farming sustainably, focusing on a different couple's farm each month. This is not just a Powell County group. Their farms are scattered across several adjacent counties, meaning that they usually have to drive some distance for the monthly get-togethers, except the host couple—the couple whose farm they focus on in a particular month. I heard about the group at a winter PFI conference, when I happened to sit down for the conference dinner at a table with two of the couples. I was fascinated with what the group was doing, and they invited me to attend their meetings. They are proud of their openness with each other, and rightly so. Inviting me to attend was part of that commitment.

In a fish-bowl culture, the normal tendency is to withdraw into secrecy, suspicion, and silence—into guarded words, image making, and petty jealousies. Rural Iowa is not Peyton Place, nor was Jane Smiley's grim portrayal of rural silence in *A Thousand Acres* meant to be taken as an empirical study.[2] But the publicness of the farmer's business, plus the stresses of the "farmer's problem" and the competition for land in industrial agriculture, have definitely driven farmers further apart from one another, literally and figuratively. These are old tensions. Still, the larger farm sizes and improved transportation of today mean that one's farm is more visible than ever before. And the continued industrialization of agriculture has every year ratcheted up the intensity of the farmer's problem and competition for land.

Let me return briefly to my conversation with Clint, the industrial farmer I talked to in Chapters 1 and 2 about competition in agriculture.

"If I had an idea," Clint explained toward the end of our interview, "and I was really smart—something that had come up saving me forty or fifty bucks an acre putting in my crops, but I'm still getting the same yield—well, I'd be an idiot to tell. You know, what I should do is just farm what I can. Rent some more ground and farm whatever I can. And don't tell anybody, until I've made whatever. Or I'll sell it," he said, meaning the idea.

That one floored me. "Right, right," I said. Big pause. "So—" Even bigger pause, while I tried to process.

"Stumped you there," Clint said with a grin.

"I think you have," I replied, and we both laughed, probably a bit too hard, not knowing what else to do.

But this kind of thinking hasn't stumped these five PFI families. They do know something else to do. Talk. To each other. About farm life, the universe, and everything.[3]

I got a sample of one of their intimate, free-flowing, wide-ranging conversations one rainy June day when the group gathered at the farm of Earl and Kerry. They raise cattle and hogs on about 350 acres, using hoop houses, pastures, and lots of crop rotation. They had recently converted half their farm to organics, and were contemplating going 100 percent organic in the next few years. In fact, that was to be the main topic of discussion for the day's meeting: Should they do it?

By the time I arrived, the group was taking a break for lunch and was chatting happily, although not without a few gloomy comments about the excessive rain of the past few weeks, which had put them all behind in their planting. It was a simple buffet lunch—bread, cold cuts, chips, and brownies. I gratefully accepted the invitation to make myself a sandwich. Earl and Kerry's is a comfortable home, I thought as I settled into my food, but not flashy. There was a new electric piano visible in their family room; otherwise the furnishings were inexpensive and somewhat worn. They have two tractors, I later found out, older models with roll bars but no cab. Earl and Kerry clearly are careful to live within their means, both in the home and on the farm.

As a newcomer, I found myself at first pretty much just listening to the waves of conversation washing across the room, paper plates on our laps.

"Neighbors aren't neighbors anymore," one man was saying to another. "They're your competitors." The other man nodded and made a reply I didn't catch.

On my other side, a woman who runs a small dairy farm with her husband was saying to the neighbor on her other side, "We could all learn from the Amish. We were led on a tour of an Amish farm by an Amish farmer. He was doing a lot of grazing. Someone in our group asked him, 'Why not put in electric fencing?' And the Amish man said, 'but I've got all these boys that need work to do.' And we never think of that."

Somehow—my notes don't say—they got from there to talking about the decline of small-town business. It's not such a big conceptual jump, in fact.

"We've got to support our local businesses. It's hard, though. I don't like paying 10 percent more."

"But we don't think of the cost of going someplace else, do we? It's expensive to drive, and the time too."

"Yeah, you've got to go thirty, forty miles sometimes now."

"We've still got a hardware store in our town. And I've been trying to shop there."

"I bet you like having him there," I put in. Probably I was leading him a bit, but I wasn't being just a scholar at that moment. We had just lost the hardware store in our own downtown, where my family liked to shop.

"Oh yeah. Because, when you need it, some tool, you need it right then. Because you've got to get back into the field. So I pay that 10 percent all right!"

Recognizing monologue is indeed a struggle, I found myself thinking, but these farmers seemed to be working hard on it. After lunch, Earl and Kerry passed out a sheaf of papers to the group. A map of the farm. A description of their enterprises. But then some unexpected things: a complete financial statement of the farm and, even more surprising, to me at least, a family goals-and-values statement that they had worked up with the input of their two teenage children. We've never produced anything like either of these in my own household. (It would be a darn good thing for us to do, though.)

But it probably wouldn't be easy. What Earl and Kerry presented to us sure didn't look easy. Nor, clearly, was it easy to present these documents, even to a supportive group like these five couples. I couldn't follow the details of the financial statement, but Earl in particular seemed to feel quite nervous about it, for, at least to him, it showed a number of significant management errors on the farm. And, indeed, there was gentle but probing discussion about several points. The family goals-and-values statement went down better. They'd done a great job there. Everyone thought that, me included.

One of the family's big concerns was Earl's workload and the high stress he often feels, which affects their whole family.

"I probably work too hard," he said. "But we all do that with our passion."

And farming is definitely a passion for Earl.

"Still," he went on, "this goes back to what Roger said this morning about what do you want." This was evidently something that came up before I arrived. "It goes back to what do you want. You know, how much do you need."

"How much is enough," added Roger, a round-faced man with a wry look.

"How much is enough," Earl said. "Yeah, that's right. And what you want to do. And what I want to do is I want to try to run a sustainable farm and pass that along on into the future, as a gift to the future. You know, maybe they'll come and see things I've done right, and the things I've done wrong, and improve on it from there. But that's what I'm going to pass along."

Everyone was quiet.

"That's what I want to do," he added, almost in a whisper.

Then Earl passed out another paper. "I know this is going to be controversial," he said. It was the plan for going 100 percent organic, mainly through converting the entire farm to a six-year crop rotation. Several of the other couples in the group were flirting with organics, but none of them had considered such a complete makeover of their operations. It turned out not to be very controversial, though. It seemed to make a lot of sense to everyone, considering the strength of the organic market, the difficulty and expense of running the farm on two systems at once, and the compatibility of organic agriculture with the environmental and spiritual concerns in Earl and Kerry's family goals-and-values statement—including their hope that a single system might ease the time demands on Earl so that he could spend more time with the kids and on household chores.

Eventually, there was a lull in the conversation. Earl was clearly rather drained by the experience of sharing these ideas with the group and getting their input. Everyone was a bit tired, in fact. And it was getting on toward 4:00 P.M., with the drive home and evening chores yet to go. It was time for some closure.

Raelyne, who along with her husband operates a roughly six-hundred-acre grain and livestock farm, caught my eye. "Well, Mike, what does the professor think about us?" she asked.

I flushed red. I was not prepared to say anything wise and professorish. "Um, I think you're great." Which was completely true. "You're working on the most important thing to grow, and that's people." Which was perhaps a bit of a gushy thing to say, but it was also completely true.

There was another pause in the conversation while I think people evaluated whether I was being sincere. Then Earl, who had been leaning forward, elbows on knees, watching me intently, sat upright and flashed a big smile.

"There's a scene," he began. "I want to tell you a story before you all go. There's a scene from *Crocodile Dundee* where they're talking to somebody who has just come back from going to a psychiatrist, and how it cost four hundred dollars an hour. And Crocodile Dundee said, 'Oh, well, when we have problems, what we do is we just tell Wally, the town barman. And then Wally tells everybody. And they aren't our problems anymore.'"

"That's right," Raelyne nodded. "It's when you keep it inside that you really have the problem."

The term I would use for what they were talking about is *response ability.* It's a bit of a bad pun, I know, but what I mean is the commitment to dialogue, to the social conditions that give to all the ability to speak and to be answered. This commitment entails *sponsoring* the words of others and encouraging the *spontaneity* of those words. That is, response ability entails encouraging others to make creative contributions to the conversation of the moment, and to the larger conversation of social life, contributions that are socially situated but also in equal measure self-willed. And response ability means doing it over and over and over again, continually renewing our sponsorship of the spontaneity of others. In fact, the etymological root of the words response and responsibility lies precisely here, in ancient testament to the importance of this social commitment: in the Latin *spondere,* which means *to sponsor.*[4]

Response ability is not a matter of polite listening, however. It is about *hearing,* and about speaking in return. To listen politely is so often merely to allow the words of another to echo into silence, unattended to. Polite listening is so often about letting others blow off steam, and comes with a not-so-subtle message of are-you-done-yet? pasted across the listener's face. Polite listening is so often a matter of disengagement, for all its pretense of engagement.

No, the best way to indicate you have heard what someone has said is to offer not a nod or an echo of their words but instead something different in return. To respond is not to parrot, nor is it to smooth feathers. We all know that others are different, that they have different histories and perspectives and interests. And that's why we seek to have dialogue with them. Because we expect them to say something different back to us. Because we want them to say something different back to us. Because we look for a life that is more than a series of monologues that we impose on others and must endure from others.

This active encouragement of dialogue is, I believe, the central wisdom of practical farming—or what we might more generally call *practical agriculture*—and what it has to offer to sustainability. It depends on understanding knowledge as a social practice, and thus on the recognition of difference as the source of sameness and change as the source of stability. That may sound paradoxical, but seeing these relationships as practical truths and not as paradoxes is at the heart of practical agriculture.

The characteristically modern way of seeing has been that knowledge is something that is true both here and there, both now and forever. The modern vision has sought permanent and universal truth, truth that is stable and everywhere the same, secured through the objective methods of the expert and verified, say some critics, through the implacable evidence of what makes money and what does not. This kind of truth comes from on high, from the absolute, and is itself absolute. This kind of truth is monologic truth—monologic both in its top-down, long-speech unidirectionality as well as in its valuing of the single logic of those who speak from the podium of modernism.

Now, no cultural orientation has been more criticized in the past forty years than the modernist one. Ever since Thomas Kuhn's 1962 *Structure of Scientific Revolutions* and Rachel Carson's *Silent Spring*, from the same year, we have been debating the consequences of modernism and the sanctity of rationalized and expert knowledge. Actually, the debate is far older—probably as old as modernism itself—but we have certainly been debating it with a renewed vigor these past forty years. And many have argued that it is time to move on to a more social understanding of knowledge and its cultivation that they suggest we call *postmodernism*.

But there are dangers here, too. Although the word postmodernism is used in a wide variety of ways, for some it has meant antimodernism, or something not far from that, based on a solipsistic understanding of knowledge, where truth is argued to be relative to one's perspective and therefore incommensurable from perspective to perspective. I know what I know, and you know what you know, and there is nothing else and no other way that either of us could know. I have my interests and you have yours. Moreover, what you or I see as true today may easily change. There is nothing to hold us to it but the shifting sands of our own perceptions. There is no sameness and stability here; all is difference and change.

Such a vision of postmodernism runs the risk of replacing one form of monologue with another—of replacing the grand monologue of modernism with multiple monologues, one for each human on the planet. If I can only know what I know and you can only know what you know, why should I pay attention to what you have to say? Why should we even bother to talk?

But there is an alternative to the multiple monologues too often characteristic of postmodernism and the universal monologue too often characteristic of modernism, and PFI farmers are discovering it. It is an alternative that embraces difference as the source of social vitality. Yet it also holds that difference is founded on sameness, for how else could we communicate or recognize our difference? And it holds that sameness is founded on difference, for how else could we communicate or recognize our sameness? To make a connection, there must be something to connect, some difference, just as to have difference there must be a social space to connect across. And as we connect these differences, we find ourselves changing, ever changing, and thus ever alive to the possibilities of the world.

Ever alive: Let me emphasize that phrase. Such an understanding is neither antimodern nor anti-postmodern. Such an understanding appreciates modernism's search for the transcendental, for a degree of sameness and stability, but sees the basis of this degree in connections and in relations, and thus in a postmodern appreciation of difference and change. It sees difference and change as essential to the central goal of life and thus of knowledge: *unfinalizability*, life and knowledge without end.[5] This is the dialogic view of stability: maintaining life and knowledge through constant growth, through sponsoring spontaneity, through cultivating creativity. And what could be more practical than that? This, then, is the dialogic faith of practical agriculture: There is only one absolute—that there are no absolutes. Or, to put it more directly, keep on talking.

But we do need some measure of common language, some measure of sameness, in order to connect across our differences and sponsor responses, as well as find a common occasion. PFI found a basis for those commonalities in 1987, two years after its founding—a novel means for conducting the "on-farm" research that has ever since been a hallmark, if not *the* hallmark, of the organization. It became almost a mantra for PFI: randomized, replicated, narrow-strip, paired comparisons. It sounds dull, perhaps, but this approach quickly proved to be a magnetic conversation starter among Iowa farmers—and, as we'll eventually come to, among university researchers as well.

"In 1985, Dick was fooling around in that same field that he had the different cropping systems in," Rick explained as we drove on from Lyle's field day.

That Dick was Dick Thompson, and that same field was one we had both visited on an earlier field day that summer. I nodded. I remembered the field. It's about as famous a field as there is on an Iowa farm, save the Field of Dreams—which, I mused to myself, Dick's sort of is as well.

"And what was he doing there? He was growing some hairy vetch. And he had, like, three practices out there. The whole field was taken up with three or four blocks." Research blocks, Rick meant, the standard technique of agronomic field experiments the world over, but much, much larger. Hairy vetch is a scraggly-looking legume, increasingly popular as a cover crop and forage crop because of its nitrogen-fixing ability and palatability for livestock. "Then, when we analyzed the numbers, we got some incredible kind of indication that hairy vetch had increased the corn yield twenty bushels, or something like that. We kind of stood back and said, 'Wait a minute. Maybe we're actually looking at the differences from one side of the field to the other.'"

In standard agronomy experiments, researchers lay out garden-sized plots to minimize the wide variability one can get across a single field owing to changes in the slope, soil, and proximity to potential influences out of the field. But plots like that are hard to integrate into the operation of a grain farm. Dick felt that if research is to be truly on-farm—if it is truly to test the practicality of an agricultural practice—then it must be part of that farm, and not just another set of university research plots in the garden of the ivory tower. Dick insisted that the research be something that was also farmable, something you could do with a tractor and a combine out in a real, live field, and without a lot of turning around of equipment and strangely laid out patterns of crop rows. A few simple subdivisions of a standard field met those criteria. But with such big plots, the problem of field variability crept in.

"So was that trial done with randomized, replicated plots?" I asked. Randomization and replication are also standard procedure in university field trials, so researchers can perform statistical tests of significance.

"There was no replication. The field was divided up into three or four different pieces, and a different practice tried in each piece."

It clearly wasn't science. Or, better put, Dick and Rick were good enough scientists to see that these results were going to be a conversational nonstarter, another mini-monologue acceptable only to a postmodern solipsist as a form of farmer's "local knowledge." But Dick and Rick are not solipsists. Just as they were struggling at the time to recognize the monologic character of university science, at least as it then was, so they were struggling to recognize the equally monologic hazard of a pure local knowledge.

Later that fall Rick happened to attend a conference where Chuck Francis, a professor of agronomy from the University of Nebraska and one of the best-known academic figures in sustainable agriculture, was giving a talk.

"I remember coming up to him after his talk," Rick explained, "and I described the problem to Chuck. And he just said 'paired comparisons.' Suddenly the lights came on."

"The problem was this hairy vetch thing?" I started to say.

"I guess the problem was how do we do something on a practical scale," Rick cut in. "Oh, I could be doing sixty-five here." A new road sign had came into view.

"How do we do research on a practical scale?" he continued, after he brought his eyes back up from the speedometer and had settled us into the new speed. "In other words, be able to farm it and still get meaningful results. In other words, get around the field variability that you would encounter when you're working on a very large scale. And Chuck's suggestion was a paired comparison, which has the advantage that you get field variability down because treatment A is always right next to treatment B. And the arithmetic is so simple that that also

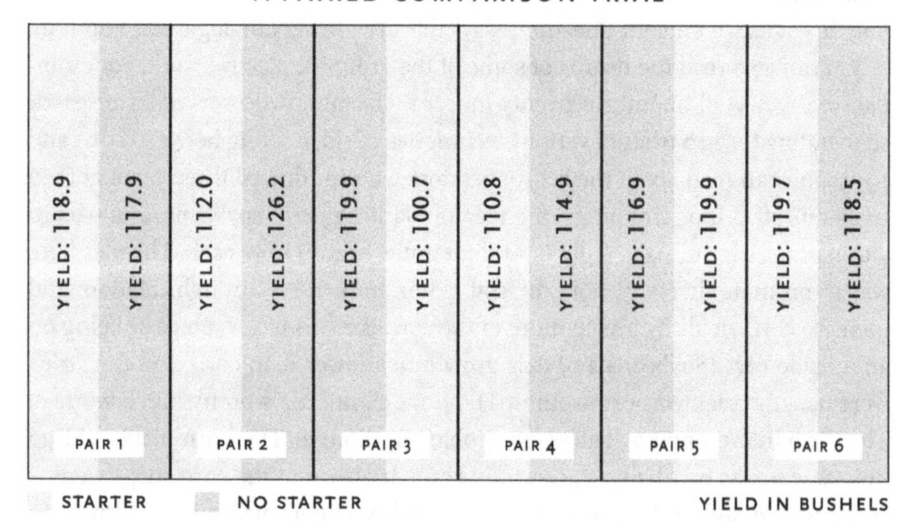

Figure 3. *PFI's narrow-strip, paired-comparisons design for on-farm research, example of a starter fertilizer trial. Source: PFI.*

makes it accessible for a farmer. . . . Then Dick added to that the long narrow strips, which let you farm it" (see figure 3).

I lost my focus on what Rick was saying as a staggeringly huge combine passed us coming the other way on the highway. "Man, those things are big. They still look big to me," I added, feeling naïve and a bit depressed by its monologic modernism.

"That one's set really wide," Rick nodded, looking at the extra-wide "head" on it, the maw into which the crop disappears as a combine crawls across a field. "So then we had a formula. . . . And until Dick got into these fancy designs, he would always do his own T-tests, you know, and I would just check his arithmetic. But basically he could do it himself, and so he could be in charge of the information-gathering process from concept to completion. Answer his own questions. And be able to have some feel for how meaningful the results were, how much confidence he could place in the results he got."

If you ever have the chance to visit PFI's web site, you can pull down the form and instructions for doing on-farm research with randomized, replicated, narrow-strip, paired comparisons.[6] There is a section on the basic logic of T-tests, sum of squares, and least significant difference. There is also a section on the logic of randomization and replication. The site gives advice on how to ensure randomization and suggests that farmers do six replications of their field trials, two more than the four replications standard among university researchers, in fact. "Replication is crucial because it gives you a 'second opinion' (and a third, fourth, . . . sixth) on your question," says the site—a very dialogic way to put it.[7]

You can also read the results of some of the group's collective work. For example, you can read about the twenty-nine PFI farmers who served as research cooperators for 140 trials of various techniques of ridge tilling between 1991 and 1995. You can read about the six cooperators who conducted three years of field trials on "strip intercropping," the idea of planting corn, soybeans, and oats in alternating strips of six to eight (and sometimes twelve) rows each. The idea here was to promote diversity, hold the soil better, improve it through rotation, and increase corn yields by giving more of the corn the sunny advantage of being on an outside row. (Soybeans and oats grow much lower than corn, and the strips were usually oriented north-south.) The ridge-tilling trials pretty much worked. The strip-intercropping had some problems, though. The corn yields did go up—way up, as much as 25 percent in some trials—but the corn strips weren't separated enough to keep corn rootworm at bay. (Rootworm hits corn crops in the second year in the same field, unless you douse the field with a soil insecticide, a practice PFI farmers try to avoid for reasons of cost and environmental impact.) Such are the vicissitudes of research. Some things work well and some things don't. That's why you do research to begin with.

And that's why PFI farmers read through the reports of the group's research trials in PFI's quarterly magazine, the *Practical Farmer*. That's why a cadre of PFI farmers shows up at the annual "cooperators' meeting" to plan the next season's research trials. That's one of the main reasons why PFI farmers say they attend the group's annual general meeting.[8] Again, it sounds dull, even improbable—farmers reading and debating tables and statistical analyses—until you realize just how practical, that is to say dialogue-based, it is. PFI farmers do.

"So between Dick's idea that the treatments, the experimental units, need to be narrow and long enough that you can drive to the other side of the field and back," Rick concluded, "and Chuck's idea about limiting the number of treatments to two so that the field variability effect is reduced and the arithmetic is simplified—you only have to do one sum of squares—we had the beginning of a good recipe."

A good recipe for dialogue.

Glen and I wandered down the hall of the Gateway Center in Ames, the hotel and conference center where PFI was holding its annual meeting the next day. It was Friday afternoon, and the pre-meeting seminar on marketing "alternative pork," pork from pigs not raised in confinement buildings, had just finished up. Actually, it had finished up almost forty-five minutes earlier, at 5:00 P.M., which is when Glen and I were to meet for our interview about his experiences as a PFI research cooperator. But it took him that long to extricate himself from the knots of conversation that had formed up and down the hallway after the seminar. PFI farmers like to talk.

"Sorry about that, Mike," he told me. "We were getting into some great arguments afterwards."

"Hey, no trouble. Now, let's see. Is this room empty?"

Yes, we could see from the window on the door, but it was locked too. The next one was both empty and unlocked. A bit spare and echoey, but it had a nice table for us to sit at and for me to set up my tape recorder.

"So I wanted to chat with you about what it's like to be a cooperator, the kind of things you've done," I began when I had the machine switched on. "Has the knowledge done anything for you? Has it not? How do the ideas come about? Who do you work with? How does it work out? Do you do the statistics? Is that meaningful? All that kind of stuff. I just asked you about fifteen questions."

Glen shot me a mock flabbergasted glare. "I was trying to answer them!" he said, emphasizing the word "trying." We both laughed. It was going to be a comfortable, friendly interview. "Sort of back up, maybe," he implored.

"So we'll start with one. You are a cooperator, yes?"

"Yes, I've had an on-again, off-again relationship. Nineteen-ninety-one would have been my first official year. Prior to that I was doing research trials on my own, following the protocol that had been developed."

"So you were doing it on your own farm?" This was a surprise to me. I didn't know of a farmer doing the trials by himself or herself, outside the larger structure of PFI. "You were following the—"

"The randomized, replicated, field-length trials."

"With six replications?"

"Yes."

"And you were doing statistics?"

"Yes. Because I had access to the paperwork." It turned out he knew some local farmers who were PFI cooperators, and they gave him the write-up of the procedures. There was no PFI website then. "We just photocopied them. Still, the statistical analysis, I know how to go through the steps to get the end result. But the background, I'm not a statistician. Don't care to be. And that's why we have other people involved, obviously."

That would be Rick Exner, PFI's longtime research coordinator, who now has a Ph.D. in agronomy from Iowa State, no less. A growing number of Iowa State faculty members are also involved—but I'm getting ahead of the story.

Glen is no stranger to technical details, however. He describes himself as "kind of a stickler for documentation or inaccuracy with numbers, much to my irritation at times." He studied agronomy as an undergraduate, and his dream at that time was to have his own farm to conduct experiments on.

"When it got all said and done, it was going to be a research farm. [My vision was of] a university field with the grass lanes and things. As the fellow that rented me my first farm said, 'Glen, if you didn't read so damn much you'd be a good farmer.'"

"Wow," I laughed. "It's a compliment, really, in a way."

"Well, it wasn't meant as one," Glen replied ruefully.

"So, would you say you trusted the PFI procedure from the beginning?"

"I would say that. Because I had seen enough of the documentation from previous years, coming to the annual meetings and things. This was what I was looking for—because there were some answers out there. I have very little tolerance for testimonials. And that's what 75 percent of the stuff being marketed in agriculture is based on."

"Right. So this is hard data you mean."

"Yes. . . . In the traditional agriculture infrastructure out there these things"—meaning research like what PFI does—"don't happen. Field trials are twenty varieties side by side. Tell someone there's a difference out there. That's absurd. Any statistician will look at that and just say that's a worthless piece of data."

"You mean that's what companies do."

"Yes. That's what's called research in the rest of the agricultural community."

A drive through rural Iowa at the height of the growing season readily shows what Glen was talking about: scattered flurries of signs on the edge of roadside fields, one every eight rows or so, identifying a company's crop varieties. "Pioneer 36R10," "Pioneer 34B24," "Pioneer 35T27"—agricultural Burma Shave signs set out to scalp the farmer, at least in Glen's view. Or pick up any recent issue of *Iowa Farm Bureau Spokesman, Iowa Farmer Today,* or *Crop Decisions,* and you'll find ads of confident-looking farmers wearing company caps, with their name, town, and a quick testimonial. Here's a recent sample:

> Barry Nieuwenhuis, Hospers IA: "I've really been pleased with the performance of all my Midwest Seed products. They dried down great. G 7706 dried 4 points in one week."[9]

Televised sporting events are also rife with testimonial ads for every kind of agricultural product. It's a nice bit of pocket money for the farmer with a roadside field or an honest face.

Vic Madsen, longtime PFI leader, combines his ridge-till corn at his farm near Audubon, Iowa, 2001.

Although he didn't put it exactly in these terms, I think that what bugs Glen about corporate testimonials is not just that they are so obviously interest-laden and self-serving. It is that they are also so monological, so undebatable. Indeed, the corporate testimonial typically seeks to appeal simultaneously through both modernist monologue and postmodernist monologue—a double whammy. On the one hand, there are the scientific-sounding crop variety names—Asgrow AG2302, Dekalb DKC57–38, Syngenta N59-Q9—and the little graphs and tables of yield trials that many of these ads feature. On the other hand, there are brief quotes from happy customers, with their names and hometowns, seeking to cultivate identification of corporate knowledge with the experiential wisdom of local folks.

If this is what advocates of local knowledge have in mind, Glen wants no part of it. Call it hometown modernism. Call it corporate postmodernism. In either case, Glen does not feel inclined to follow the conversation, probably because he recognizes that these conversational positions are fundamentally structured by money, and thus not conducive to close scrutiny and debate. Nor are they open to change, to the unfinalizability of knowledge.

Glen did, however, get closely involved in a debate with Dick Thompson and some other PFI farmers about a nutrient stratification problem some of them were having in their fields after several years of ridge tilling. The potassium largely disappeared from the root zone in the soil profile. So they tried broadcasting potassium (laying it down over the whole field); they tried banding potassium (laying it down on the rows only); they tried injecting potassium (using a special planter attachment to put potassium below the seeds). None of these ideas worked particularly well. But eventually they hit upon a solution that, given the widespread (and justifiable) concern about soil erosion in Iowa, at first seemed radical: plowing in the spring, once in a five-year crop rotation. Figuring that their use of cover crops, grassed waterways, buffer strips, and hay rotations put their soil loss well below the norm for Iowa farms, Glen, Dick, and a few others decided that it was worth the risk to do an occasional careful plowing, immediately followed by planting (as opposed to the fall plowing that is still common on some Iowa farms).

"Work me through the knowledge generation, if you don't mind," I said to Glen. "Had you gotten together at a cooperator's meeting? Where everybody looked around and said, no, this doesn't work?" What I meant was that broadcasting, banding, and injecting didn't work.

"We just looked at our data."

"And you were reading through the newsletter? Or how would that exchange have taken place?"

"I'm sure that between Dick and I, we had a lot of phone conversations during that era. And of course when we'd get together for the winter [PFI] meeting. . . . And then we get on the phone if we feel the need and say, 'Why did that work there and not here?'"

After many years of research trials, Glen eventually made the decision to go organic.

"And I probably wouldn't have taken that step if I hadn't had the knowledge to be able to document what I was doing, and some people there to hold my hand."

"So not just the randomized, replicated plot design, but you mean other ways of documenting what it is that you're doing?" I was confused. "What do you mean there?"

"Just sitting down with people and arguing about how do we structure our costs and things."

Arguing: Midwestern farmers arguing with each other, and feeling good about each other afterward.

"It has been very valuable. In my organic operation now, I have weed control issues using the flame weeding and rotary hoeing and things." Like the flame cultivator, the rotary hoe is another of the standard weed-control tools for organic grain farmers. It's a series of barbed disks that crumble the surface of the ground, disturbing the contact between the soil and the roots of young weeds. "How the flamer fits in. Actually that was my PFI trial this year."

"Okay."

"I had a statistically significant difference in weeds and yield. So I was pretty pleased with the trial."

"Using the flamer versus the rotary hoe?"

He nodded. "In my mind, the flamer is the last tool in the tool box for early season weed control. Ridge till first, then the rotary hoeing, and some harrowing depending on the situation. I choose to use the flamer last because I think it is the most severe to the crop. There are people that argue that point with me, and that's next year's trial."

Not only are randomized, replicated, narrow-strip field trials a good recipe for dialogue, for response ability, between farmers. They also invite the participation of university scientists. In 1987, the year PFI implemented their field-trial recipe, an Iowa State University extension entomologist, Professor Jerry DeWitt, took up the invitation—or rather invited himself—to one of Dick Thompson's field days.

"That was the field day where Jerry DeWitt showed up as we were ready to go out in the wagons after lunch," Rick recalled, as he drove on. "Jerry showed up in a three-piece suit and Dick got out of the bucket."

I looked puzzled.

"They would lift Dick up in a bucket loader so he could be up there fifteen feet up in the air talking," Rick explained. "I would hold the PA for him, and he would talk."

"Got a picture of that?" I laughed, trying to imagine the scene.

"Quite a few pictures. So Dick got out of the bucket, talked to Jerry for about a minute, came back. He was quite pleased. DeWitt had gotten back in his car and left, and Dick said 'Sounds like we're going to work together.' And so that began—"

"This was in the middle of the field day?"

"Yes."

"Wow."

"DeWitt didn't stay around," Rick chuckled, perhaps thinking about the three-piece suit in a crowd of jeans and feed caps.

"So this was at Jerry's initiative?"

"Jerry really gets the credit for this, yes."

Jerry assigns the credit the other way around. We discussed his involvement in PFI one afternoon in his campus office, his desk, table, and floor piled high with papers, extension reports, sustainable agriculture brochures, and unopened mail, gently composting away. We talked excitedly for an hour about a wide range of topics concerning PFI. At the end, he beckoned me closer with a wave of his hand and said quietly, "PFI has meant a lot to me personally. I would not be who I am today—I would not be where I am today or doing what I'm doing today—if it weren't for PFI. It's been that important. Sometimes you look back and you can see events that were real turning points in your life that you might not have realized were at the time. Well, PFI has been that for me."

Who Jerry is today is one of sustainable agriculture's most ardent and articulate advocates on a university campus. He also works closely with a division of the U.S. Department of Agriculture called the Sustainable Agriculture Research and Education Program.[10] Jerry describes himself as having begun his career as an "uncomfortable traditionalist." But by 1987 there was more than lack of comfort. He was a shaken man, ideologically, trying to come to grips with the farm crisis and his growing feeling that the university's supposed high-tech vision of agriculture was blind to the real needs of farmers.

"The university lost a lot of innocence in the farm crisis," Jerry explained to me. "We thought we had all the answers and we discovered that we didn't."

Indeed, the farm crisis gave the university a whopper of a black eye, in the minds of many. It is no secret anymore that the research priorities of our universities are heavily influenced by where the money comes from. Particularly in an age of declining state and federal support, public universities increasingly find themselves in a nonstop scramble for "sponsored research"—code words for external, often corporate, dollars. For example, the state of Iowa's contribution to Iowa State's budget fell from about 36.3 percent in 1995 to about 32.7 percent in 2002.[11] This may not sound like a big drop, but in a nearly $1 billion annual budget it means about $35 million less. The money has to come from somewhere, making it hard to turn down offers of corporate-sponsored research. Moreover, the grants that are available from the federal government increasingly bear the hallmarks of corporate pressure in the allowable categories for funding.[12] And in order to retain legislative support for the remaining funds from the state of Iowa, the pro-business mood of the day requires that the College of Agriculture pay close attention to the wishes of industrial agricultural groups like the Iowa Pork Producers Association, the Iowa Soybean Growers Association, Iowa Farm Bureau, and the like. As a result, say some, the university is more responsive to the needs of the economy, and that's a good thing. Others, however, argue that being "pro-business" really means being pro-*big* business, to the detriment of independent family farms and rural communities. The influence of big business on the university is an old trend, of course. But the 1980s farm crisis, combined with a rising consciousness about the environmental consequences of Big Ag, brought these issues into public discussion like never before.

And not just into public discussion. Some in the university, like Jerry DeWitt, came to question the one-sidedness of the research that was being done. So DeWitt, who was associate dean of extension at the time, brokered an unusual collaboration between Iowa State and PFI, impressed as he was with the group's commitment to scientific research and dialogue. PFI's work was growing, and it clearly needed to hire at least one staff person. DeWitt arranged that the university would provide the group's staff person with a small office on campus, and would also consider the staff person to be on the university's extension staff as an "extension associate" and would make the university's benefits system available to that person. PFI would pay the staff person's salary and the cost of his or her benefits, and PFI's decision making would remain with its farmer board of directors. The university did, however, retain an equal say with PFI in the hiring of the extension associate. But there was little question in anyone's mind that this person would be Rick Exner. The university also helped out in the early funding of the extension associate position by giving PFI a grant to work on energy efficiency in agriculture, for which the university had itself been given a pot of money.[13]

Jerry traced the conversation between PFI and the university back to a point before Dick climbed out of the bucket loader to meet Jerry in his three-piece suit. The Plant Pathology Department at Iowa State hosted a seminar series on the farm crisis and invited Dick Thompson to give one of the seminars.

"It was the most formal I have ever seen Dick Thompson. I do remember that Dick Thompson was extremely nervous. And that was probably equal to the anxiety in the room of what was this guy going to say, and what was he going to do. So it was sort of a moment of, like, two dogs looking at each other and not quite knowing what the other was gonna do, or who was gonna move first. It was sort of a stalemate. He was nervous, we were nervous. It was a quietly electric moment in Iowa State's history."

But the talk went well, and Jerry was impressed, very impressed. Like many of the other professors in the audience, he expected Dick to "run down the university," as many farmers at the time were doing. (Many still do, perhaps with some justice.) But there was none of that in Dick's presentation. Jerry heard Dick's main message loud and clear: Farmers and university researchers have a lot to talk about.

In the weeks that followed, Jerry got to thinking about how to take the conversation forward. "It was just, how do you get started after all of these misconceptions? After all of what people think, how do you actually get it going? . . . PFI felt strongly about on-farm research. And they wanted to test and to make observations on their farms. The university has their model of laboratory and field testing on small plots. There was a tension [in my mind] concerning, 'Why isn't PFI, farmers' collective data, valid and viewed as a resource by the university?' The university general response was—you could never get anybody to basically say this publicly—but, well, it was on the farm, [so] it wasn't valid. PFI was saying, 'Well we have been working with others from around the country. We do these comparisons. We do the statistical tests. Why isn't it valid?'"

In other words, things were still very much at the two-dogs stage.

"The major event that I recall that began to turn that one around," Jerry continued, "of all places—you will enjoy this greatly—we met at McDonald's. Rick Exner, Dick Thompson, and me, and I believe it was Steve Barnhart," another professor from the university. "There were four of us. I remember it was very uncomfortable."

"Who had set this meeting up?" I asked.

"As I recall, it was most likely Rick Exner on behalf of PFI, and me, saying, 'Well let's sit and talk about their plan for on-farm research.'" Jerry reached for a swig of his can of pop. "Their question was, 'If we presented a plan to you on how we conduct on-farm research, can the university react to this? How can they improve the plan if they don't like this? How can we get them to recognize that our research is valid and does contribute?' That was a very tense meeting. We went in there not knowing if we were going to argue or not. They said, 'We've got something we think you ought to recognize.' And I represented, in a sense of body, [the view] that probably was saying, 'Well, what you're doing is not valid.' Across that little table there was a lot of tension."

"So at that point, they weren't asking for university collaboration?"

"No. They just wanted us to recognize that they were valid."

So Jerry went back to campus with a little diagram of randomized, replicated, field-length paired comparisons that Rick and Dick had given him. "At that point, I'd just started my tenure as associate dean. So I was in a position to take it to people and say, 'What do you think?'"

"Right."

"And I remember in particular that I shared it with Reggie Voss," another professor of agronomy at Iowa State. "I said, 'Here's what they'd like to do. Here's how they will do it. With side-by-side comparisons. It'll be randomized. They'll run this statistical test on it. What do you think?' Reggie looked it over and said, 'Yep, that'll work. That's valid.' Poof! When I heard Reggie Voss say, 'Yes, that's valid, that'll work, that's fine,' it was like, now wait a minute. For how many years have we been discounting what they're doing as not workable? And they have been thinking we will never recognize their work. It took one meeting, an hour meeting, a piece of paper, and a why-don't-you-look-at-it. It took me ten minutes to give it to somebody and Dr. Voss to simply look at it, and poof! All of that tension was over with."

So Jerry got in his university car and drove out in his three-piece suit to Dick Thompson's field day and shook his hand.

"I look back at that meeting at McDonald's, and the diagram shoved across the table, and that changed how we viewed their research. That changed how they viewed our attitude. It was a big event. And to me," Jerry searched for the right words. "Mike, I'm saying, all it took was a brief meeting and one of our senior scientists to say, yeah, it looks good. I look back at that, and the question is, not that it was approved. The question is, why in the hell did it take so long for minds like Dick Thompson and Reggie Voss and others to get together and agree? They were agreeing. But we never got together to ask the right questions to agree. Why did it take so long?"

Better late than never, one of the great guiding principles of the social commitment to dialogue. Dialogic providence works out that way sometimes, even usually—and that's if it works out at all. Again, dialogic providence is not mere chance. It doesn't mean that a group like PFI can only sit back and hope to stumble into the right associate dean of extension at the right moment in time. It doesn't mean that PFI can only hope that some farmer in the midst of a phenomenological rupture will suddenly decide to attend a PFI meeting. PFI and its members can try to start the conversations of providence. They are more likely to be well received if they start these conversations on a dialogic footing—if they use a language that invites the participation of all potential speakers, through encouraging their response ability rather than turning them off with monologue. A language like randomized, replicated, narrow-strip, paired comparisons.

"Science with practice," reads the motto on the insignia of the Iowa State University of Science and Technology, the full handle of the institution. "Guess that means we're still practicing to be scientists," goes the standard joke on the ISU campus. Which, all joking aside, is exactly right: That's what science really is about. Science is not about final truths, much as we have often been taught otherwise. Science is an ongoing process, an unfinalizable conversation we have with others and with the world. The most dedicated scientist is the one who most deeply hopes that the results of her or his work are speedily made out of date by new work, new evidence, new debates and dialogues. No, science is not about absolutes and universals and permanent truths, and neither is farming. Both science and farming are about taking what we think we know and putting it into practice, ever encouraging, ever listening for, ever sponsoring the responses of others and the world, and in turn offering back our own responses.

Practice, thus understood, does not make perfect. That is not its point. Perfection is another vision of the absolute, and a sense that there is an endpoint, that we can reach it, and perhaps that we have already reached it and that there is nothing left to say. But to reach the end of a conversation, and to shut off the possibility of future conversations, is a kind of death, a monologic death. Rather, the point of practice is to keep us alive, to keep us growing, changing, creating. In other words, the point of practice is sustainability—in our farming, in our science, in our dialogue with life. Farming with practice, science with practice, dialogue with practice, life with practice: Together these are the maxims of practical agriculture.

8. New Farms, New Selves

The grain fields and county towns have been whizzing by for a couple of hours now. I stopped in the last town for self-serve coffee and a home-made doughnut, 25 cents for the coffee and 25 cents for the doughnut. (Really. Good, too.) So I'm juiced up for another hour of driving to another meeting of that group of PFI couples who get together every month to talk about each other's farms.

The towns up here, two counties north of Powell, are particularly small and struggling, well removed from Iowa's few urban centers. I come to the next town. There's the traditional welcome sign on the outskirts, with a list on it of every local business, church, and club—a kind of one-page yellow pages for the town. I slow to thirty miles per hour, as requested. In among the plywood storefronts, I'm pleased to see several old businesses still going, as well as a couple that seem to be expanding. Big Foot Hardware-Grocery. First Security Bank and Trust Company. Pete and Shorty's Lumberyard. Doc's Bar and Grill—Cold Beer, No Extra Charge. There's a Catholic church and a United Church of Christ, both in immaculate shape.[1]

This month's meeting is at the farm of Owen, a new member of the group. Owen's not married and doesn't have a partner right now, and so represents something of a broadening of the group's definition as a group for farm couples. He's in the process of taking over his parents' 450-acre grain and livestock farm. The purpose of the meeting is to discuss his plan to convert from a standard two-year corn and soybeans rotation to a six-year rotation with two years of pasture, two years of alfalfa hay, and a year each of corn and beans. To make it work, he'll have to switch from bringing the feed in to the cattle to bringing the cattle to the feed—that is, to grazing them on the pasture ground, plus supplemental feeding with his own grain. He's planning to use intensive grazing methods, meaning that he'll be rotating the cattle from paddock to paddock on a nearly daily basis. It's a big change to consider.

But the bigger change for Owen may be the meeting itself. He sounded pretty nervous about it when I called for directions. As I pull up the gravel drive to the Victorian farm house where he was raised and which his parents have recently vacated for him, Owen is talking under a big shade tree with several of the other

Vic Madsen, longtime PFI *leader, at his farm near Audubon, Iowa, 2001.*

men from the group. I get the impression that they are trying to get him to relax a bit, or at least to let him know what to expect.

"We don't hold anything back," I hear Earl saying, as I walk up to them. "Hi Mike," he nods toward me. "We let each other have it, don't we Mike?"

He turns back to Owen. "And when you come down to my place, I'm going to get it back from you."

"They really put me through the wringer," says another of the men, Roger.

I'm not sure it helps. Owen still looks stiff.

Everyone's arrived now, so we head inside to Owen's living room. He has filled out the circle of living room furniture with some kitchen chairs. It's a bit crowded, but there are seats for everyone.

The session begins with each member of the group saying a few words in turn about what is on his or her mind. Roger starts, perhaps still thinking about the conversation under the tree.

"You've got to get yourself out of the box. That's what we're trying to do here. Like when we went over my corn sheet last time, and I took some bruises." Evidently, Roger's had been the farm under the microscope last month. "You know, you feel like a fish out of water, flopping around. And this group helped me through that a bit."

There are nods all around at this, including from Owen. The conversation continues around the room. My attention drifts, as I try to decide what I'm going to say when it's my turn. (Even though it's my second visit to one of their meetings, I'm still feeling a little awkward as an outsider in such an intimate group.) In any event, there seems to me a sudden break in the flow when Cheryl grabs all of our attentions, out of turn. She just has to speak.

"I just want to say," she says, speaking loudly at first, but then quietly, and looking right at Roger, "what you said about being a fish out of water. That was a hard thing, especially for a man, to say. That says a lot about what's good about this group. That we can say these things."

Especially for a man: Cheryl was pointing to something fundamental. Sustainable agriculture involves more than agricultural practice. It is about the practice of the self—or, better put, it is about the practice of selves, new selves, new men and new women and new dialogues between and among them. It is about new conceptions of the unfinished wholes within which we all live. It is about new cultivations of farming, and that means both new senses of what a farm is and can be and new senses of what a person is and can be.

Given agriculture's patriarchal patterns in Iowa, and most everywhere else, changing farm men's knowledge cultivations is crucial to the sustainable cultivation of practical agriculture—to the dialogues necessary for a practical agriculture. That's a rather over-packed and academic sentence, I know. So let me try again. As powerless as farm men may feel themselves to be, as much as they may often feel like fish out of water, flopping around, the cultivation of Big Ag encourages them to suppress this thought and to exert what control they can, and in the process, often, to suppress others, other people and the land. Big Ag encourages them to conceive of their masculinity as a kind of monologue, as the independent tough guy who is always right, always in charge, always too short on time to attend to the needs of others, even if it may sometimes pain him. To the extent that men are successful in exerting this control, and to the extent that the social and economic structures of rural life support their control, men are in the tractor seat of agricultural change. And in most tractors, there is only one seat.

There is an alternative, however, and many PFI men are discovering it. In place of a monologic understanding of masculinity, there is sustainable agriculture's commitment to openness, to the unfinalizable, to admitting one's mistakes and dependencies, and to encouraging other voices, including that of women and the land. This commitment resonates with a different kind of masculinity—a more *dialogic masculinity,* a masculinity that talks, a masculinity that comes down from the tractor seat.[2]

Dialogic masculinity is also something that many PFI women are discovering. They are discovering it in the greater consideration and acknowledgment given to their words and interests and contributions. And they are discovering it as something that they can help create, for gender is a relational matter. Gender is something we do, in interaction with others, both male and female. Gender is something we practice.[3] PFI women are part of those relations of social practice, and thus have some leeway to encourage change.

In so doing, PFI women change themselves. For sustainable agriculture is not a matter of dialogic masculinity alone. It is a matter of dialogic femininity, too. There is more than one party to any monologue, and many farm women have had to accommodate themselves to a place in the audience, as well as to a simultaneous place backstage handling the props and scenery of the show. For all its pretense to independence and individualism, a monologue is a social performance—a social practice. Dialogue is, too, but one in which we are all speakers, audiences, and stagehands for each other. Dialogue thus requires that men and women have a different sense of femininity as much as it requires that they have a different sense of masculinity.

But let me not speak only in terms of dialogic masculinity and dialogic femininity. Let me not bound things such. Let me speak rather of dialogic *people,* and of the history of their encouragement in PFI. For PFI has changed a lot as an organization, and so too have the people involved in it. When it began in the 1980s, PFI focused almost exclusively on issues of on-farm research oriented toward the needs of grain and large-animal livestock farmers. Much of the excitement was about trials of techniques like ridge tilling.

As Glen told me, speaking about the early years of the group, "At that time PFI was somewhat inaccurately labeled as a 'ridge till club.'"

Only somewhat inaccurately, Glen admits. It was also somewhat of a ridge till *men's* club, for grain, cattle, and hogs are overwhelmingly male enterprises on most Iowa farms, including most PFI farms. The leading figures from the early years of the organization were all men. Dick Thompson, Rick Exner, Ron Rosmann, Tom Frantzen, Vic Madsen. It was not until 1996 that PFI appointed its first female board member. And while there were usually more women in attendance at its activities than at those of most Iowa farm organizations—it is still not unusual for there to be no women at all at many public farming functions in rural Iowa—men and men's voices were far more prominent in PFI's early years.

But PFI was an organization born of a commitment to dialogue, and a commitment to dialogue has the wonderful quality—yes, the wonderful quality—of encouraging still more dialogue, ever widening its embrace of the diversity of the world. Since the mid-1990s PFI has greatly broadened its mission and the voices it celebrates. It would now be highly inaccurate to call it a ridge till club, or a ridge till men's club. The organization has seen a big shift of focus toward a more *relational agriculture.*

Like any new trend, a new lingo has come with this shift: community-supported agriculture (CSA), direct marketing, relationship marketing, institutional buying, community food security, food systems, local foods, slow food, multifunctional agriculture, agritourism, and more.[4] But there is a lot behind this lingo, as well as the increased importance of older phrases like farmers' markets and marketing cooperatives. Among other things, it represents a great increase in the participation and prominence of PFI women on their farms and in the organization itself. Almost all of these new activities involve women far more centrally than grain farming and large-animal livestock farming do. Many PFI farms now raise vegetables, herbs, small fruits, chickens, lambs, goats, eggs, and sometimes dairy products, with an eye toward local markets. Instead of raising bulk commodities and selling them through central buyers to customers they never meet—albeit raising them on smaller farms and in more environmentally sensitive ways—PFI farms are working hard to establish direct relations not just with customers but with communities of eaters. PFI women have made the majority of these connections, and are doing much, if not most, of the raising of these nonbulk products. The organization now includes several women on its staff and on its governing board. There is an annual PFI "women's winter gathering." And in 2002 the board appointed PFI's first woman president, a full-time CSA and goat farmer, Susan Zacharakis-Jutz.

A few numbers may help highlight the difference PFI's dialogic commitment has made in the organization's gender balance. As of 2003, three of the eleven members of PFI's board of directors were women. Also in 2003, twenty-two of the sixty-six workshop presenters at the group's annual meeting were women. Thus women represent a quarter to a third of the group's leadership, which is no paragon of gender equity but quite high compared with the large, commodity-oriented farm organizations in Iowa. I haven't conducted a systematic survey on this point, but here are a few examples. As of 2003 there was one woman among the fourteen members of the board of directors of the Iowa Farm Bureau, the "State Chair-*man*—Farm Bureau Women," (my emphasis), a dedicated seat. There was one woman among the seventeen directors of the Iowa Corn Promotion Board, and one woman among the seventeen directors of the Iowa Soybean Promotion Board. There were no women at all on the board of the aptly named Iowa Cattlemen's Association, although there were several women on the association's staff. Happily, in 2003 there were two women among the eighteen directors of the Iowa Pork Producers Association. But then the Iowa Pork Producers Association, it should be noted, still hosts an annual "Iowa Pork Queen" and "Iowa Pork Princess" contest, with the winners chosen from among the County Pork Queens every year at the annual Iowa Pork Congress. (A contestant for a county pork queen competition—twenty-two Iowa counties still hold them—must be a high school junior or senior, and she must be from a pork-producing family. Pork producer organizations in some other states, such as Wisconsin, run similar annual contests.)

While at the time of this writing men remain more prominent on the whole, at least in terms of officers and speakers at meetings, PFI cannot now be considered an organization for men. Rather, it is an organization for people, dialogic people, seeking a more practical agriculture for their earth.

I have bordered on tossing to the wind a couple of well-established social scientific cautions in the previous section. Let me say at least a few words about those cautions. First, it is not among the usual reflexes of the social scientist to make the kind of normative statement I did concerning the "wonderful quality" of dialogue. But despite the extensive coverage I have given to the struggles and uncertainties of contemporary agriculture, I believe there is cause for some optimism in agriculture. Optimism requires a moral benchmark. So here is mine: Dialogue is good.

It is important to keep my moral benchmark in view because of a second social scientific caution: that we should be mindful not to confuse the way we would like the world to be with how the world actually is. What I am trying to argue here is that the commitment to dialogue that characterized PFI from the start—that characterized the on-farm research trials and the effort to engage the university and other farmers—contributed in time to a more inclusive dialogue. For once the issue has been raised that someone else could be, and needs to be, part of a dialogue, the issue of that next someone is immediately implied. Thus PFI rippled out from its beginnings, embracing in time the concerns of women, family, and the eaters of what farmers grow. And I believe PFI's dialogue is rippling out still further yet. The social scientific problem is that I might be sifting the evidence of our fieldwork to show this—to show it both to readers and to myself—because I would very much like it to be true.

But all research faces this problem, even the most supposedly neutral. Every investigator desires to have a result, thus making every research result a kind of moral good, susceptible to the sifting and resifting of an unruly world into an unrealistic coherence. Rather than sweep it aside with the false rhetoric of objective detachment, I believe the best way to handle this issue is to be up front about our moral commitments. In this way, we can have a better sense of what to watch out for as we consider an academic argument, listening critically for what has been left out and left unsaid, so as to broaden the argument's own potential dialogue. This dialogic broadening should be everyone's response ability, and I hope I have contributed to that ability by alerting readers to my moral inclinations, as well as reminding myself of how I might be jiggling the sifter.

Elinor and her husband Frank farm about three hundred acres in western Powell County, with a fifty-head cow-calf herd of Black Angus cattle, a small herd of sheep, a CSA vegetable garden, and a few-dozen-per-day free-range egg operation. Most of their land is pasture, hay, or woods. They raise only forty acres of grain, all "open-pollinated" corn, which is corn whose seed you can save and perform your own selection trials on. The per-acre yield is lower than with hybrid corn, but after the first year you don't have to buy any more seed, and you can work on creating varieties that fit the local conditions of your own farm. Also, open-pollinated corn sometimes resists pests and disease better than hybrid corn, most of which has been selected mainly with yield and chemical inputs in mind. Consequently, the economic and ecological yield with open-pollinated corn can be higher—particularly when you use it as an "internal output" that you feed to your own livestock, as Elinor and Frank do, rather than sell it as a bulk commodity. In fact, everything their livestock eats comes from their farm. Elinor and Frank sell no bulk commodities, only food, marketing almost all of it directly to those who will be eating it. Nor do they use farm chemicals or genetically modified crop varieties, although theirs is not a certified organic farm.

But when Elinor and Frank began farming in the early 1990s, taking over Frank's late grandmother's land, the farm was pretty much a standard corn-and-beans setup, plus cattle, albeit with more woods than most Iowa grain farms have. And when they began, Elinor had nothing to do with the daily operating of the farm, although she was bringing in almost all of the family income through her job as a technical writer.

"I think Frank came back to it and discovered something in himself that he had put aside for a long time," she said, as we enjoyed tea around their sunny breakfast table. "And that was the satisfaction of tilling the soil and putting something in and taking care of it and getting something out of it. That was as good a match as anything for him to do."

"And for you?"

"I was a few years behind the curve on that. Because for me, that was just something he did. And I got the benefit of living in a really nice, natural setting. As long as I could bring in enough cash to keep the whole thing going, that was fine. But I could do completely different work."

While Frank had a farm background, Elinor did not. She grew up in the suburbs.

"So what kind of farm did you grow up on?" I joked.

"Exactly. No farm background whatsoever. We went down sometime each May to Alfredo's, or Denardo's, or some Italian name who had the nursery. We bought the pansies and the geraniums and brought them home. That's it. That's what I knew. You go to the nursery, you get them, and stick them in the ground."

"So you didn't really have an inkling back then that you'd ever wind up doing something like this."

"No, not a clue. Isn't life fun?"

But while Elinor is still the "cash cow" of the farm, as she puts it, still bringing in about half their income and—most important, she said—securing the family's health plan, she is now, ten years later, very involved in the whole operation.

"I've just grown more and more into being more and more an integral part of the farm, as opposed to really just the cash cow. Ever increasingly. I'm still the cash cow, but I'm more and more involved in the decisions and the planning, and in fact thinking up some of the things that we should try to do."

"You mean with the field work and the cattle as well?"

"Just a lot more in the decisions. Yes, the cattle too. The whole direct marketing thing. I think Frank always knew it was there. But it took my effort to really make those connections start snapping. Yes, we'd sell a quarter," meaning a quarter of a beef cow, about a hundred pounds of meat, one of the standard sizes of a direct marketing order. "If someone called us, you're like, 'Yes!' And somewhere down the line a couple of months later we'd get it together and get a calf and call this person back. And hopefully they were still interested. Now it's much more of a business."

While she is involved in the field work and the cattle, Elinor is primarily responsible for the free-range eggs and the CSA vegetables they grow. These are barely minimum-wage operations, when Elinor and Frank calculate the income from them separately. But Elinor sees the main financial return from the eggs and vegetables as the contact they establish with the local community, which then leads to sales of the beef and the lamb.

"We're a livestock farm. But vegetables are such a great entrée. The eggs are such a great entrée. People—"

"You're trying to sell your meat through the CSA," I interjected, trying to see if I was tracking with her.

She nodded. "People who don't know us and love us aren't going to necessarily slap down the initial hundred dollars or whatever you need to buy your first quarter of beef. But they'll give you two dollars and try the eggs. That's not a big transaction. Then they go, 'Wow, these eggs taste really different than the ones that I buy at HyVee.'" (HyVee is one of the local supermarket chains.) "'And they're much better, and they're keeping much longer in the refrigerator. Hmm. Those eggs were great.' 'You know, we have beef and lamb too.' . . . I'm going to give them more excuse to come onto this farm. I'm actually going to be in dialogue with them more."

A pickup drove into the yard, and Elinor immediately reached for a dozen from the small stack of egg cartons from the day's laying. It was a retired male cousin of Frank's, coming to drop off some free lumber he thought they might be able to use.

"Just a minute," Elinor said.

But it was about fifteen, with the screen door half open. A bit of family gossip. A bit of politics. A bit of joking and commiserating.

"Isn't that just like rural Iowa?" Elinor said, coming back into the kitchen as Frank's cousin finally returned to his truck, with instructions on where to dump the lumber and a free carton of eggs under his arm. "Fifteen minutes at the back door of the mud room, but he wouldn't come in." He did leave some free lumber, though, plus something else that those eggs bring to their farm: community.

According to Elinor, PFI had a big role in bringing the eggs to their farm, and what the eggs bring in turn. Frank had been participating in the group's activities for years, going to Dick Thompson's field days, going to the annual meetings, reading the results of the research trials in the group's newsletters. Sometimes their travels in the state would take them relatively close to Dick Thompson's farm.

"He'd go, 'There's Dick Thompson's.' I'd be, like, 'Yeah, Dick Thompson's.' He started going to PFI field days. Frank definitely hooked us up with PFI long before I was paying attention to them."

What made the change was an opportunity to do some volunteering for PFI, and Elinor's eventual discovery of some PFI women who were becoming more involved in their farms.

"It put me constantly in touch with people who were doing what my husband was doing. I really started to think more and more about what he was doing."

But she also starting thinking more about PFI women, such as Cindy Madsen, who has a one-person poultry operation and direct markets her chickens and much of the pork and lamb her husband Vic raises. And one of us, Donna Bauer, who had become a PFI board member.

"Cindy Madsen, it was like, actually I'm kind of like her. I think I'll pay attention to her. Oh, Donna Bauer, I think I'll pay attention to her. Suddenly the women I was communicating with and constantly in touch with vis-à-vis PFI turned out to be these women that I could have a lot in common with, if I pursued those threads."

"So they were kind of mentors in a way?"

"I wouldn't say that they were mentors. I would say that I sort of began—"

"Models?" I suggested.

"I found a tribe, I would say. I found a tribe of women who were interested in sustainable agriculture. 'Oh, it's not just this thing. There's lots of people interested in this.'"

By "tribe" I think Elinor meant an identity, a cultivation of knowledge and her own self—new social relations of knowledge and a new phenomenology for her and her farm. Suddenly she found herself with new things to say, and new ways of saying it.

"Frank would read, talk, engage me in what he was doing," Elinor continued, before she began farming herself. "So I wasn't out of the loop. But it was like, 'Yeah, yeah. Whatever you want to do, honey. You want to put in some alfalfa over there? Okay.' I would even enjoy helping putting it in. I like being outside. But it was like, 'Yeah, okay. Yeah, five acres, seven, I don't care.' But now it was like, 'Hey, look what I read today. This guy inter-seeded alfalfa with rye,' or whatever. 'Have you ever heard of somebody doing this?' . . . I said, 'We could try this.' Frank said, 'We sure could.'"

"Don't answer this question if you don't want to." I reached for the tape recorder, making a silent offer of shutting it off (as we had earlier done for another part of the interview). "Did Frank resist?"

"Like, to turn some of our valuable acres into vegetables?"

"No, did he resist you getting more involved in the decision making?"

"Oh no. I think he welcomed it. I think it was a huge burden for him. For all the joys, I think it's a really scary business to be in. I'm happy to answer that question because it's only done better things for our marriage to be more involved in the farm."

This is relational agriculture in the most personal sense of the term.

I interviewed another relational farmer just a few days before I spoke with Elinor. Noah has been running a two-acre organic CSA garden since 1997. He grows a huge range of vegetables: carrots, radishes, cucumbers, beans, peas, potatoes, tomatoes, peppers (hot and sweet), half a dozen lettuce varieties, easily a dozen kinds of salad greens other than lettuce—a living smorgasbord from the catalogs of Johnny's Selected Seeds, Seeds of Change, and Seed Savers Exchange.[5] But he's probably best known in his neighborhood for his luscious raspberries: red ones, yellow ones, black ones. He started with eighteen families in his CSA and now he's up to forty, which is plenty, he says. It's already, with that many, his full-time job for six months of the year.

Noah was raised in rural Iowa, but he isn't a farmer's son. His father had grown up, in Noah's words, as "a dirt farmer" in a poor rural area, but had left all that behind before Noah was born. Or mostly behind. His dad's work was professional employment in town, but the family lived in the countryside. So Noah had a lot of interest in rural matters growing up, and he had a lot of friends who were farmers' kids. Noah knew something of what Iowa corn-and-beans agriculture was all about. And he knew he didn't have any interest in it.

"I spent a lot of time out in the corn fields. I spent a lot of time walking beans," he reminisced at his kitchen table with me, steaming mugs of mint tea from his garden in front of us. "Walking beans" means pulling truculent weeds by hand from the soybean fields. It's part of the childhood of most rural Iowa kids Noah's age, which is about forty-five. "But I really didn't have any interest in actually doing the farming myself. My dad's not mechanically inclined, and I didn't grow up that way. I just don't like dealing with big machinery. As it turns out, as it evolved for me, this wasn't something I was thinking about. I don't know how, innately I kind of knew that."

Noah was deeply interested in agriculture, though, and got his undergraduate degree in agronomy. But he didn't know what to do with it. He thought about going to graduate school, as his father wanted him to do. But "I told my dad, you know, I think there's enough information already. It's just not getting into the hands of the people who can really use it."

So Noah joined the Peace Corps, working for several years in Africa, trying to get that information into those hands. As for so many PFI farmers, living overseas was a transformative experience—a phenomenological break from life as he had conceived it to be.

"I got off the plane in Africa and kissed the ground because all these years I'd been waiting to do this. Everything was in the future and now I knew—I was very conscious that I knew that now is the time for action, and now is the time to just really be present to the moment and to what I'm doing. It was a wonderful experience, an opportunity in working in grassroots development with very poor people, and living right with them, right out there in their homes. We had our own separate home, but a house with a tin roof. That also gave me a deeper appreciation for what my father grew up in. I think everybody should have that experience. We could all live that way quite well, actually."

"And you found the folks down there were as happy as they are up here," I added, probably leading him a bit too much.

Noah nodded. "But I also realized, I don't have a lot of practical experience."

Moreover, during those years in Africa, he cultivated an appreciation for hand labor, so crucial to the cultivation of his CSA garden.

"Working overseas, all the work that those farmers were doing was by hand. I'm just more of a hands-on person in that sense and would rather work with my hands rather than with machinery. So it kind of evolved that way for myself."

But when Noah returned from his development work overseas, he still saw no place for himself in agriculture. He was married now, to Carol, whom he had met in Africa, where she too was a Peace Corps worker. They had a young son, and they were casting about for what to do next. So they both worked as organizers for a variety of causes for several years—peace work and lay ministry for their church (they're both deeply spiritual people) in the main. But the organizing work eventually drew to a close. Noah was looking for a new opening in his life, a new focus.

"I was beginning to ponder, okay, what are we going to do next? It was before our CSA got started, and I hadn't heard about CSAS or any of that. But I was feeling like I really need to be connected somewhere or another in agriculture again. Two different people, who didn't probably even know one another, told me, 'You know, you ought to go up to this Practical Farmers of Iowa conference. That might be a good place for you to connect.' So after the second one, I said 'Okay. I'm doing it.'"

It was an occasion that resolved his phenomenological break, so long left flapping, with the agriculture he had grown up amid. Not only did he eventually learn about CSAS through PFI. This meeting brought him into contact with another form of cultivation, so crucial to his current garden: that of practical, dialogic, relational agriculture.

Angela Tedesco, PFI *member and a community-supported agriculture (*CSA*) farmer, with a tray of vegetables at her farm near Des Moines, Iowa, 1998.*

"You don't find people in the States that are asking the right questions very often. You may find a lot of people that resist, or they raise a lot of questions. But they don't have hope. It was really tangible."

"That's very true, what you say about that, the sense of overwhelmedness and cynicism and withdrawal," I added, in honest support.

Noah gave a vigorous nod. "PFI was the first place I'd been—this had been like five years after we'd been back—was the first place I'd ever been that I got a sense of that. A sense of, they could critically think. They could critically analyze what was going on, politically, culturally, environmentally around them."

Critical. Exactly, I thought.

"They could put that together, and they were trying to do something positive. They were looking for solutions. So it was forward moving."

Practical. Yes. I nodded vigorously.

"It was hopeful. It didn't mean that they were idealistic, or that they had it all together."

Unfinalizability.

"But that here was a group of people that was positive. They knew how to celebrate."

Cultivation and community. This too is central to the dialogic vision of practical agriculture. Noah is directly involved in the cultivation of practical agriculture through on-farm research trials in his garden, as a PFI research cooperator. He's particularly interested in nonchemical control of a nematode problem in his carrots, and in the use of a backpack version of a flame cultivator. (Noah's is an organic CSA garden.) He is involved in exchanging the results of that information through field days, phone calls with other growers, and through good times at meetings and community celebrations for sustainable farmers. In those critical dialogues, in that social flame cultivation, built upon a commitment to response ability in which difference is an invitation to further engagement and mutual cultivation, Noah found an agricultural home that was accepting of his own difference.

For Noah, I found, is an unusual man in many ways. He is not just a farmer. He's the family cook, and he does most of the housework during the winter months, when he's not farming. In fact, he got into farming *so* he could do more housework. He found that with his organizing work, he was away too much.

"I was away from my two-year-old son. Carol was over-time working as homemaker, which isn't her cup of tea," he explained.

I knew he was a baker, as I had a piece of his pie once at a community event for Iowa CSAs. "I've had your pie," I said.

"I do most of the cooking."

"I think it was your cherry pie."

"Not cherry. I don't do cherry."

"Maybe it was rhubarb."

"Yes."

There is some remaining gender asymmetry here. For most women, doing at least an even share of the housework, or even a greater share of the housework, is not generally a source of personal pride and community praise. It is usually assumed that a woman will contribute at least that much to the daily tasks of home life. And although there are more and more heterosexual couples where men do contribute an even or substantial share of home labor, it remains a source of some pride and praise for them (and pride of their partners in them)—in some social circles, at any rate. Noah evidently felt that our conversation that afternoon constituted one of those circles, and he felt able to indulge himself in some pride in telling me about his pie making and all the cooking, cleaning, and child care he does. Quite justifiably, in my view.

His different sense of himself as a man is also reflected in the kind of farming he does. Small CSAs like he runs are associated in many men's minds as not real farming, or perhaps as not real men's farming. I told him a little story to get at this.

"I had a student once who went out to visit a CSA farm in our area. I won't tell you the name of what the farm was. I should say 'CSA garden' because, at least from the student's point of view—he was an Iowa boy from a big grain, big tractor background—he came back and he said, 'You know, it wasn't a farm.' Would you say you're a farmer?"

"People are telling me I'm a farmer. That's been my experience. I know a lot of farmers in Africa that two and a half acres is what they farm, all by hand. I till the soil. I work the soil. I plant seeds and I harvest and care for the earth."

"Is that the sort of thing a man should be doing?" I said with a smile. He laughed. He knew exactly what I was talking about.

"You'd have to understand when they did that little study about the women in CSAs in Iowa, that—"

"Betty Wells and Shelly Gradwell." Betty Wells is a professor of sociology at Iowa State. Shelly Gradwell is a former graduate student there.

"That Betty Wells and Shelly Gradwell did," Noah continued, nodding. "I read that. I got a copy of it before it was done and I said, 'That's me.' I fit the descriptions of what they were going through, child care and all these different issues. I'm not a typical person, a typical male that way at all."

(That paper, later published in the journal *Agriculture and Human Values,* is entitled "Gender and Resource Management: Community Supported Agriculture as Caring Practice." Well worth reading.)[6]

Some farmers, Noah went on to explain, "Well, they're all into equipment, machinery. Anything you can do with a machine they'll do it. Put this together. Put that together. Really into that kind of thing. I can't relate to that. I can't relate to that at all. That is meaningless to me. That's a real guy thing, usually. Carol, my wife, would be better at that. She's more of a fix-it, mechanical person."

Carol doesn't work on the farm much, though. She's "in charge of the herbs," is how Noah put it. Carol's main paid work is in town, about twenty hours a week as an administrator at a nearby community college. But they both strive to maintain roughly half-time jobs, hers half-time each week, and his half-time each year. I got the sense that she and Noah do their best to contribute roughly equally to the household economy, both in cash work—Noah said that they each bring home about $12,000 a year, for about a thousand hours of labor each, with Carol's job also providing the family health care benefits—and in non-cash work. But I got the sense that they also have a kind of too-many-cooks view of trying to coordinate their actions in the same work on the farm, hour after hour. I can't be certain here. I was not able to interview Carol. She happened to be away for a few days when I visited their farm. But my impression is that, in the main, theirs is an equality of labor with a division of labor. This is a peace they have come to.

Whether or not I am right here, peace is certainly Noah's larger goal.

"If you hadn't heard of community-supported agriculture," I asked him, "do you think you'd be doing this?"

"Absolutely not. From my faith perspective, it was a work of the spirit, the timing behind everything. . . . As far as my faith aspect of what I'm doing, I've realized that this is how I work for peace. It's hands on. It's something practical. It's something positive. It builds community. It's nurturing. It gives people life. With my work as an organizer, we did a lot of activism and we did a lot of talking. We organized rallies and did a lot of things, but it was always working against someone else's agenda. We talked about your limits." Knowing your limits was a topic from earlier in the interview. "I've come to understand I can't do everything. I can't solve the world's problems. But at least here in this small place, I can do a small bit that can give life, to my family, to all my family, and to those few people that I touch through this endeavor. . . . I think that's very much a fundamental part of what I'm doing here."

"Trying to make peace through farming."

"Yes. Considering how destructive farming has been and how competitive it is. You talked about this change." Again, this was an earlier topic, but actually from before the interview began, when we chatted about the general decline in rural community. "And I think that's where, in my opinion, we need to shift the paradigm from competitive to cooperative. Food is not about scarcity. It's about abundance. If you work from that paradigm, then everyone's needs are met."

"Right."

"And everyone benefits."

Farming for us all.

The new selves associated with the new farms of practical, dialogic agriculture go well beyond issues of gender. Every self is a self in context. The dialogic self, however, is a self that is encouraging of and responsive to its contextual and relational origins in all their many varieties. Or, better put, to the degree that a self is so encouraging and responsive, that self is more dialogic. PFI farmers, men or women, are not perfectly dialogic. No one is. But in their best moments, they show a strong commitment to dialogic interaction with a broad range of others and their differences, as well as a critical eye for the structures that so often stand in the way of dialogic interaction: monologic structures of power and inequality.

I say "in their best moments" because dialogue is still often a struggle for PFI members. No, they're not perfect about it. I'm not either. None of us is. And the circumstances of PFI farmers still often discourage dialogue and cloud their critical understanding of monologic power. But PFI farmers are committed to trying, and they do often succeed, sometimes wonderfully.

For the rest of this chapter, let us sample something of the range of these dialogic struggles and accomplishments. Let us visit PFI farmers from across Powell County and Iowa, listening in while they talk about family, admitting mistakes, tolerance, critical thinking, politics, spirituality, globalization, corporate power, materialist desire, productionist desire, animals, and the environment. Let us listen as PFI farmers look for the roots of sustainability in the response ability and unfinalizability of new selves and new farms.

"Yeah," Sue agreed, early on in her interview with Ryan, who farms about eight hundred acres organically, along with his wife and in cooperation with another couple. "In a similar vein, what important changes have you made to your farm in recent years that you would see as important?"

"In recent years, I think the biggest thing has been kind of a shift of priorities," Ryan replied. "I grew up on a farm where the farm was first. Everything else came second. I saw the conflict in that once we had a family. The family really was something that was awfully important. There was a time when I struggled a lot with that. And with the last few years, I think there's been a shift of priorities more toward family being the most important. Farming sort of has to come to second. There are times, you know, when you're planting and things, like where, say, you know this is important," meaning the farm operation. "But you know, I've got, especially with our family, with our kids, emotional problems and stuff. Our philosophy was you deal with things when the problem comes out. And there have been days when I spent half a day holding, counseling kids, when you knew you should have been doing something," again meaning with the farm operation. "And that was hard. That was really hard, especially to start with. But once you do that and realize the value of that and see the rewards, then it makes it a little easier. So, in that way, that's probably the biggest change—just a shift of priorities away from making the farm the one and only, all encompassing."

"That's great," inserted Sue, supportively.

"Because, you know, the farm is not eternity. It'll come and it'll go."

Family and unfinalizability.

"Those three P's are just like an addiction of anything else," continued Dale, referring to his little mental shorthand for paradigms, peer pressure, and prejudice that we discussed in Chapter 7. "Well, with drugs or whatever, the first thing in order to overcome an addiction is that you've got to admit to it."

"Right, right," I said.

"And it's just like me admitting that those things controlled my life. And then, you know, you've got to come to terms with that and say, 'Yeah, they did.' And I have to admit that. And you know I've done it to myself, and to God. But you know I've come to the point where I'll admit that to you. That I've made these mistakes. I've made these errors, and they've made me better."

Dialogic openness.

"Do you consider yourself Republican, Democrat, or an Independent, or is there some other political group with which you identify?" asked Greg of Fred, who farms about four hundred acres, mainly pasture for his cows and sheep, and mainly for direct marketing. (Greg was reading from the questionnaire we followed loosely in many of the interviews, particularly the early ones.)

"Used to think of myself as a liberal Republican," replied Fred with a yawn. (It was late in the interview.) "Now I kind of consider myself a conservative Democrat, I guess. But I've flirted with the Libertarian Party for a few years, so I don't know. I think I'm becoming more liberal in my old age in some respects. A more—definitely more tolerant. I think tolerant is becoming a real big word to me, anymore," following a Midwestern idiom in which "anymore" can mean "these days." "Tolerance of other ideas, of other people. Other ways of doing things. You might not always think it's right, but it's still there. It kind of comes into the idea that, I guess, that there's no way that I could ever feel that I was so right about doing something my way that it was, you know, that that was the only way. I never feel that I know that much, now, that my way has got to be the right way, you know. There's got to be other ways of doing it that can be just as good."

Response ability.

"What, in your view, are the most important issues facing farming today?" Sue asked of Meryle and Delmar, going back to our loose questionnaire. "Some of these you may feel like you've already touched enough."

"I really think it's getting it to the next generation," began Delmar, referring to the issue of family succession on family farms.

"That, and keeping it local," put in Meryle.

"That's going to be a trick," said Sue.

"Keeping it local," repeated Meryle, with quiet emphasis.

"And, if we can't keep it local, I don't see how we can keep it—give it to the next generation," Delmar added, intertwining his earlier thought with Meryle's.

"That's right," Meryle agreed. "And my, you know, I keep saying this over and over again, but when you get the production of the food in the hands of a few, you're going to get in trouble. Because what if that few get a disease, or get a breakdown in that?"

"That's a scary thought. I hadn't even thought about the disease," said Sue, nodding. Mad cow disease. Contamination of processed meat and fast food by virulent strains of E. coli and *Listeria.* Salmonella on mangoes imported from Brazil and on California almonds exported to Canada. Salmonella in a national brand of ice cream; 224,000 people in forty-eight states fell ill over that one, back in 1994.[7] Every year in the United States, with its reputation for the safest food supply in the world, 76 million people come down with a food-borne illness. About 325,000 wind up in the hospital because of it, and 5,000 die.[8] It is indeed scary stuff.

"And if a few could control the food," continued Meryle, "they can control the people. And that to me is much more scarier than not having any food."

Recognizing monologue.

It was after my afternoon haying with Dale and his son, Jeff, when we were sitting around the dinner table, wolfing down hamburgers that Yvonne, Dale's wife, had prepared. They were interested in my work as a professor, and Yvonne asked me how I lecture.

"Actually, I try not to lecture," I replied. "Really, the thing I try to do most is to teach students to think for themselves, and to think critically. What I try to do is to teach people how to teach themselves. I believe that's the most important thing we should try to teach." And I went on to explain my view that lecturing isn't the best way to teach critical thinking, and that university education across the country is increasingly stressing a more interactive teaching style. I didn't use the phrase dialogic education, but that's certainly what, as a fan of the writings of Paulo Freire, I had in mind.[9]

"Yeah, that's a real conservative approach to teaching," Dale said, when I had finished my little explanation.

This was, to say the least, a perplexing comment to a left-liberal academic type like myself. Dialogic education is something I had always seen as leftist, even radical, given its description in Freire's *Pedagogy of the Oppressed,* long a leftist touchstone. I thought I'd misheard him, and I think a blank look must have come over my face.

"Yeah," continued Dale, seeing my confusion, "and the other stuff, that's the liberal approach. Where they just tell you things, and then you have to kind of follow along. They really lead you around by the nose, instead of teaching you to think for yourself."

I didn't know what to say, and I imagine I still looked blank. I think Dale must have decided I didn't agree, and he seemed to be seeking some little way to externalize his views, and thereby smooth things over a little bit, when he added, with a little laugh, "As you can tell, I listen to Rush Limbaugh a lot."

The struggle to recognize monologue.

Wendell and Terri have been debating what to do with their hog operation. They have recently switched to ridge tilling and a five-year rotation in their crop ground, raising more grass and less grain. But they're stuck on what to do with the hogs. Like many Iowa farms, they have a small confinement building, built in the 1970s. It's on nothing like the scale of the vast facilities built under contract since the 1990s, though—the ones that have been driving so many farms out of the hog business. Wendell and Terri are worried that they may lose their hog business too, unless they can find another way to raise them, and maybe also a way that is associated with a niche market. Besides, Terri is simply against raising hogs in confinement.

"Wendell was saying that you're very interested in the pasture farrowing," I mentioned to Terri. Pasture farrowing is a way of raising pigs out of doors, on grass instead of concrete. Some 150 Iowa farmers have recently developed a marketing cooperative that sells this kind of pork across the country, calling it "natural pork."[10] Wendell had told me that Terri wanted to join in.

"Yes!" she enthusiastically replied. "I'm really trying to push that, trying to get the hogs out on pasture. I don't believe in the confinement. I've been watching it. I don't like it."

"What don't you like about it?"

"Well, it's not healthy for the boys," Terri said, meaning their two young sons. "It's not healthy for the hogs. It's loud, it's dirty, and the hogs don't like it. We were just out there the other night, penning up a sow, and I thought it was so stupid.

Because he was bringing the sows with little pigs out of the confinement building, and throwing the new ones in to farrow. I said, 'Why don't you just put up stalls and let them farrow out here? It'd be just as good.' But it just doesn't work out that well."

She thought a moment.

"I'll just have to take over and do it myself."

The struggle to establish dialogue.

I was at Glen and Julie's annual PFI field day. There was a good crowd, perhaps twenty-five in all, come to look at Glen and Julie's rotational grazing and to hear the results of their on-farm research on soil fertility and profitability. They had conducted the work in cooperation with a researcher from Iowa State University, and the researcher was giving his part of the field day spiel. Glen held up a series of charts as the researcher talked about them, arguing that farmers shouldn't look at yield but rather at their return per dollar invested.

"If your corn is looking dark green right through to the end of the year, you're throwing your money away," he told the crowd, meaning that if your corn is dark green that long you've spent more on nitrogen than you'll get back in yield, however good the crop may look.

Earl was at the talk too. As the presentation was going on, he came over to me. I was standing at the edge of the circle of folks listening to the presentation, having found a cool spot in the shade of Glen and Julie's John Deere.

"All this fertility stuff must be a bit boring for you," Earl said to me, in a kind of classroom whisper between naughty pupils. Earl knew by then that I was a sociologist and that I didn't have a farm background.

"Well," I replied, a bit uncertainly, not knowing where this was going, "there's a lot of it I can't follow."

"Besides," said Earl, "it's all social. That's where the real change has to come from. All this, this is just technical."

The practical necessity of dialogue.

I was in the area, interviewing another Powell County farmer, so I thought I'd drop in on Elwin, as I'd warned him I might do that day. He greeted me with his usual broad smile when I walked into his farm office, where he'd told me I might find him.

"I'm just looking for some overheads here," he said. "I've got to make a presentation about weed control at a conference tomorrow."

Like many PFI farmers, Elwin is forever getting calls to present the results of PFI's on-farm research at conferences. This particular conference turned out to be called something like "The CRP and the Future of Family Farming," CRP being the Conservation Reserve Program, which pays farmers to take erosion-sensitive land out of production for a ten-year period. Since organic certification requires that the land be pesticide-free for at least three years, many Iowa farmers have tried to start up organic production on former CRP land, as these fields come out of their ten-year set-asides, bypassing the usual three-year transition period.

Elwin eventually found part of what he was looking for—a table on how ridge tilling without herbicides has the same yields, over twenty-nine PFI trials, as ridge tilling with herbicides, and at a far lower cost. We began by talking about that and other technical topics, but eventually our conversation began to widen.

"It's embarrassing," Elwin said. "It's hard to talk about the other stuff, the soft stuff. It's easy to say what I need to do is go out and change the size of the sweeps on my plow, or go get a new auger. Stuff like that. That's easy and that's what we farmers mostly talk about. Not this softer side of things."

He thought a moment. "We're a quiet people."

He thought a moment more. "I guess I see this as PFI's main role. When farmers first come to us, usually they're interested in the technical and economic side of things. Our job as an organization is to support them in broadening out their thinking to the softer, social side."

The practical necessity of a dialogic self.

It was one of those interviews where the interviewer just has to sit back and let the words of the interviewee flow. Greg and Derrell hit it off from the start, and there was no stopping the conversation, even after Greg ran out of recording tape three hours into the interview. Derrell had a bad accident a number of years ago and had to give up the manual side of farm work. He continues to manage his acres, but he hires out the field work. Maybe it was the accident. Maybe it was his passion for writers who criticize the conventional, industrial, bigger-is-better way. (John Kenneth Galbraith is his personal favorite.) Maybe it was due to a bit of both, plus more that Greg didn't hear about, despite the length of their interview. But Derrell had strong doubts about the transcendental satisfactions of material success and materialist ambitions.

"It seems to me what you were talking about before," said Greg, a good hour into the interview, "with some of the problems you see with community because of technology, that maybe sustainable people are making sacrifices, and maybe trying to do some community things that may do some good in the long run."

"Yeah," agreed Derrell. "I think those type of lifestyles are gonna be so much more open to the compassionate, humane society Galbraith was talking about. So that when people do need assistance, of whatever kind, those are the people I can see it coming from. Because they're gonna be able to help. I think they're gonna be able to help more effectively because they've walked in those same shoes. They're living with limited material resources, and yet in many ways they're living more enjoyable, fuller lives, and their quality of life is quite good. And I do think we've got a society—a significant part of our society—of people that are desperate. People don't have a spiritual life. They're not satisfied with their job. Their marriage isn't going the greatest. They're disappointed with their kids. And yet they don't really know what the problem is. They don't know, they don't seem to be aware of, maybe there's another way, a different type of way to earn a living. Maybe there's a different type of spending pattern. Maybe there's a different type of choice of things. 'I don't need a big house. I don't need three cars. I don't need that boat. I don't need all this stuff that just cost me insurance and everything else.' But that's what they grew up with. That's what they were taught. That's the way they were trained to think financially, and they just don't see the alternatives."

Greg nodded and kept listening.

"And so they're living that desperate, unexamined life. Don't know what they're here for. Don't know why they're here. The only thing they know is that they're terribly dissatisfied with their life. They're very discouraged, very disappointed. And they've got themselves in this acquisition thing, this materialistic society that's just gotta have. You know, like a dog chasing its tail. It's no good. They don't like it. They're harassed, they're tense, they're anxious. They're all that, but they don't have any training or background or very little experience or encouragement from anybody around, that maybe there's a different way of doing this. Maybe there's a different way of approaching this life. Not that we have the answers," meaning PFI and sustainable farmers. "*You* are the only one that's really gonna have the final answers. The ones you come up with, they're going to be the satisfactory ones. But maybe we can help you."

Derrell laughed lightly.

"Help you in your process as you figure out the answers that are going to be unique to you," he continued, "and your very individual being. They're not gonna be good for me. I can't tell you what to do, 'cause that ain't gonna work for you, what works for me."

Greg gave a chuckle of understanding.

"But you need to find something that's gonna work for you," Derrell went on. "And then that's gonna keep changing on you. That's gonna be the tricky part. Once you got to thinkin' you got it figured out, you're finding out, 'No, I haven't figured out a damn thing.' It's all changed."

Difference and unfinalizability.

"Uh, what, in your view, are the most important issues facing America and the world today?" asked Sue, later on in their interview.

"Oh," said Ryan, thinking. "America *and* the world?"

"Together or separately, as you see appropriate," Sue clarified.

"That comes down to my basic belief in sin and salvation," Ryan responded. "And the most important issue has always been, and will always be, man's reconciliation to God. That's really the only basic issue facing anybody. So, I guess for me, it all boils down to that. And one of the ways that evil seems to be playing itself out, in our Western world, would be [the idea] that, you know, salvation comes through having your own way. Being isolated, being alienated from people. Boy, you get your own thing, and you're independent, and you consume as much as you can. Consumption is going to, you know, make you happy. Stuff all these things in your Dodge-sized hole that only God can fill. Stuff all this stuff in there. And because we have all this stuff, we can just keep doing and doing that. But it's not helping anything. And then there's two-thirds of the world that would just like to eat. And it's very hard for us to have any idea of what that kind of life is like. So, I think we kind of see this in America, as in the world at large. The rich getting richer and the poor getting poorer, kind of thing."

Materialism, inequality, and monologue.

I was at another of Dick Thompson's field days. (I try to go every year.) Dick was leading the crowd on to the next stop, and I got to talking with Orval, who, like me, was hanging in the back. Orval raises hoop-house hogs on a farm about the same size as the Thompsons' place. I had seen him at several other field days but had never had a chance to visit with him. Dick had mentioned at the previous stop that the Thompson farm makes about $110 an acre profit, despite refusing federal subsidies, and Orval and I were trying to sort out whether that figure included the money from the livestock part of the farm, or whether that was only from the crops.

"Well, it's really hard to calculate what farm income really is," Orval said. "But they're blue collar people, like all of us. Folks can tell that."

That intrigued me, and I said, "Really? Are farmers blue collar or white collar? I mean, you're wearing a blue shirt today!"

Which he was. Orval laughed at that, and then said, shortly, "Blue collar."

"Not white? I mean, there's that old distinction between those who own and those who owe."

"That's good!" Orval said, with another laugh. But he didn't offer an answer. I think he was trying to sort out where farmers lay in relation to this phrase.

"I guess farmers are a bit of both," I said, trying to sort it out for myself.

"Well, I don't know," Orval responded, with a bit of a drawl. "Yeah, I think own and owe. That fits. Because some own and some owe, and some both."

He paused, and then added, "And us in PFI, we're just trying to farm without taking away from others. Trying to farm small."

Materialism, equality, and dialogue.

Francis Thicke is one of the best-known sustainable farmers in Iowa, and a prominent member of PFI. He lives some three hours from Ames, just outside Fairfield, Iowa. Finally, I bit the bullet and drove down for one of his field days.

I was glad I did. Francis's farm is amazing. He is making a full-time living off a fifty-cow dairy herd, small by contemporary standards, using organic and rotational grazing techniques. He does buy some barley for the cows to eat as he milks them—not corn, having gotten worried about pollen drift from genetically modified corn. But other than that, his cows eat what Francis grows for them on the farm: grass, just grass, plus other pasture plants like clover and chickory. (Actually, it could be said that in rotational grazing the cows grow the grass for themselves, fertilizing with their manure and harvesting with their mouths.) They're an unusual breed of cows for Iowa, small brown Jerseys instead of the usual big black-and-white Holsteins. Francis uses the rich Jersey milk for something even more unusual: the only on-farm milk-bottling plant in the entire state of Iowa. Radiance Dairy is his label for the milk, yogurt, cheese, cream, and sometimes ice cream that he makes from what his cows make from all that grass. You can't get his milk in very many places, because fifty cows don't produce all that much milk, particularly Jerseys—which is why most dairy farmers don't use them anymore. But that's okay with Francis. He doesn't measure how well he's doing by how much he's producing.

After Francis led the crowd of seventy-five or so through his pastures, showing us some of his on-farm research trials, he brought us all back to the barn for some farm-made strawberry ice cream. Delicious. Then he led us inside to his dairy parlor, where the cows get milked twice a day.

I was particularly eager to see this. Francis uses "New Zealand style" or "swing-over" milking machinery. Most dairy parlors on Iowa farms take twelve to sixteen cows at a time, with one "cluster" of milk pumps for each cow, plus a monitoring system to measure the output of each cow, usually resulting in a tangle of hoses and pipes and wires and, in newer parlors, digital readouts too. But it's an expensive tangle, a very expensive tangle. In a swing-over system, there is one cluster for every two cows in the parlor, hanging in a line down the center. You hook up the clusters to the cows on one side, while you bring in the cows for the other side and wash their udders. About the time the second side of cows is ready for milking, the first side is done, and you "swing over" the clusters to side two. Half the clusters, half the hoses and pipes and wires, half the cost. Or even less, at least in Francis's case, as we discovered, for Francis has also dispensed with the monitoring equipment.

I'm rather proud of my cow moo, and gave throat to one as we filed into the parlor. Francis, it turned out, was right behind me.

"Hey, that sounds pretty good," he said. I reddened and cut my moo short. Francis smiled and finished my moo for me, getting a nice echo off the parlor walls and concrete floor. I think he'd done it before.

It was a bit crowded in the parlor, after we finished filing in. We all found room somehow. Francis began to point out the various features of his swing-over system, including its lack of monitoring equipment.

"So Francis, how do you monitor your cows' production?" I asked. "Do you do a once-a-month check or something?"

"I don't monitor them. I don't need to. To be honest, I don't really care much about production."

A couple of people in the crowd looked at each other.

"That's going to take care of itself if I treat my cows right," Francis continued. "It's just not a big concern of mine. When I look at a cow, I'm thinking about a lot of other things. Whether she calves and breeds easily. Is she easy to work with. Is she an efficient grazer. Stuff like that. I mean, it's not like I don't care at all about production. And I do have a fair idea within maybe 10 or 20 percent of how a cow's doing. I can tell by looking at her udder, and how she's milking. But there's so much more to it than that. Most farmers, production is all they think about."

"So how old is your oldest cow?" someone asked.

"Right now, I think fourteen. Pretty soon she'll be ready to retire."

This is very old. Most dairy farms rarely keep a cow older than five or six, or even older than three or four on some of the large dairies with a thousand cows or more. Cows tend to decline in milk production after a few years. Then there is the period just before the birth of a new calf, when every cow's milk production goes down. Usually farmers "dry off" their cows during this period, taking them out of the milking cycle and sending them to pasture. On some big dairy farms, however, the cows are milked straight through to their next lactation instead of being dried off. These are the cows that don't make it to four.

"'Retire.' That's great. Love it," someone behind me said to a companion. And Francis really does mean retire. Instead of sending an old cow off for cat food, Francis finds local people who will take her as a family cow or as a pet. The Fairfield landscape is consequently dotted with Francis's retired cows, living out their days in quiet munching. And his own farm is dotted with cows "on vacation," which is what Francis calls drying off, plus the working cows, out for a good feed in the pasture grass. On many dairy farms today, the working cows rarely see the out-of-doors.

"'On vacation,'" repeated that same person to the same companion, after Francis used the phrase. "Outstanding."

Bringing animals into the dialogue of practical agriculture.

As I came walking into the dining room of the StarLite Inn in Ames, where PFI usually holds its annual cooperators' meeting for planning the year's on-farm trials, I noticed that there was an empty seat next to Earl. I plunked my plate from the buffet line down on the table, and took the seat. It's always fun and interesting to talk to Earl. We had a great time, chatting away about the state of farming today, debating whether there was a new farm crisis or whether the old one never stopped, talking about the importance and satisfactions of joint decision making in a couple, arguing about whether PFI should become more political and lobby directly on farm legislation, and musing over what it really is that makes a sustainable farm tick.

It was the last topic that seemed to most hold our joint interest. I told him, jokingly, that I had a tongue-in-check, one-question test of whether someone is a sustainable farmer.

"I ask someone if they're a sustainable farmer. If they say yes, then I know they're not. If they say no, then I know they are. Sustainable farmers are never satisfied with where they're at. They're always still tinkering. So they never think they're sustainable."

Earl laughed. He liked it. I didn't use the word unfinalizability, and he probably would have been puzzled by the word, at least initially, if I had. But he understood the concept. (And he also understood that I don't really have such a test.) Yet it was somehow clear that he thought there was a deeper dimension to dialogic unfinalizability than continual learning, although he didn't put words to it just then.

After dinner, and after the after-dinner activities, I came back to the table to collect my coat to go home. Earl was still sitting there, nursing a can of beer, alone, which was surprising for such a gregarious man. He looked up at me.

"You know, I love my cows. I think that's what really makes the system work."

Bringing animals into the response ability of practical agriculture.

The weather was broiling hot in that humid, midsummer, Iowa way on the day of Wade's PFI field day. When I arrived, the gathering crowd was seeking shelter under an enormous white oak in the front yard of Wade's farmhouse. In fact, the tree was the topic of discussion as I walked up. Wade was telling someone that he figured the big oak was a prairie savanna tree from before settlement, and that his house was located where it was precisely because of that tree.

I noticed something in the grass. A wrench. I picked it up and held it out to Wade. "Been missing this a while?" I lightly taunted.

He laughed, and so did several others in the crowd. "This is my shop, out here under the tree," he explained, smiling.

Then I noticed a screwdriver in the grass too, and picked that up as well. Everyone laughed again, Wade included.

"I had a job one summer collecting fossils in the Big Horn Basin of Wyoming," I said, which was true. It was a great job. "And I got very good at finding things on the ground." Which was also true, plus I hoped it would soften any embarrassment he might be feeling.

"We've got some old fossils around here," said Wade, unperturbed. "I remember once seeing some snails in the rock down by Lake Powell. That rock's three hundred million years old. And right next to it in the water were the exact same snails, alive."

"Well," objected somebody in the little cluster of folks listening to our conversation, "they couldn't be the same snails after three hundred million years."

"Okay. Nearly the same," countered Wade. "Still, the thing that gets me is that these creatures have been around almost unchanged for three hundred million years. And who are we to come around doing the things we do? Whose world is it, anyway?"

Bringing the environment into the response ability of dialogic, practical agriculture.

"What would you say a good farmer is?" I asked.

"Define good," Bert replied.

"I think that's what he's asking you," said Shelly.

"Yeah, I want you to define that," I said to Bert, with a nod of thanks to Shelly. "What would you call a good farmer? Do you ever think of some people as good farmers, and some people as not being good farmers?"

"I think of some people as doing some things that are destructive."

"Yeah. Right."

"Or doing some things that are inventive and constructive."

"Yeah."

"I kind of hate to tar anybody, you know."

"Yeah. Yeah."

"Respect for land, animals, and society."

"Yeah."

"Probably cover a lot of it."

Indeed.

Bringing everyone and everything into the response ability of dialogic, practical agriculture.

Intermezzo

There is a word that appears throughout this book, a good, direct, sensible word: practical. There is another word that appears just as much—another good, direct, sensible word: dialogue. They often appear on the same pages, the same paragraphs, even the same sentences. I would like now to put them (or rather versions of them) together into the name of the same theoretical concept: *dialogic pragmatism*. These words, I believe, aptly describe the lived theory that underlies the struggles and accomplishments of the Practical Farmers of Iowa—a merging of two philosophical and sociological orientations, pragmatism and dialogics.

Pragmatism is the better known of the two, so let me begin with this deceptively simple idea. "What is pragmatism? Perhaps I can save time for some readers by giving the answer: no one really knows." So writes David Hildebrand at the beginning of his book that seeks to answer this question.[1] And that's because it has been a very fertile idea since Charles S. Peirce propounded the "pragmatic maxim" in an 1878 paper, leading to the writings of William James, John Dewey, and George Herbert Mead, and then in more recent years to the pragmatist revival brought on most notably by Hilary Putnam and Richard Rorty. The argument of the pragmatic maxim is that truth and meaning lie not in the perfection of method, or in eternal essences, or in the wisdom of authority or tradition, but in practical consequences for everyday life. Here's how Peirce put it: "Consider what effects, which might conceivably have practical bearings, we conceive the object of our conceptions to have. Then our conception of these effects is the whole of our conception of the object."[2]

James, pragmatically enough, called it simply the "cash-value" of an idea.[3] Dewey similarly wrote that "truth is a character which belongs to a meaning so far as tested through action that carries it to successful completion. . . . From this point of view, verification and truth are two names for the same thing."[4]

Pragmatism has often been criticized as being an idealist or even solipsistic philosophy by those who feel it allows anyone to say anything about anything, depending on the perspective of their own practical experience. It has also been criticized as a materialist or universalist philosophy by those who feel its proof-is-in-the-pudding view gives the pudding authority to dictate truth, ignoring that our perception of what is "the pudding" may be clouded by culture or perspective. Dewey's response to both of these sorts of criticism was his emphasis on process and on testing and retesting in knowledge. In this way, said Dewey, we may achieve the best truth we may achieve, what he termed "warranted assertibility"—not truth in an eternal or absolute sense. Crucial here is the possibility of open criticism from others. In his words, "free thought itself, free inquiry, is crippled and finally paralyzed by suppression of free communication. Such communication includes the right and responsibility of submitting every idea and belief to the severest criticism. It is less important that we all believe alike than that we all inquire freely and put at the disposal of one another such glimpses as we may obtain of the truth for which we are in search."[5]

But how do we communicate that warranted assertibility and that criticism to others? Does this not require us to represent the world to others? Does not this need for representation open up all claims about the world to corruption by power? If so, then pragmatism is open to postmodernism's deep critique of language and representation. Much recent pragmatist thought has centered on Rorty's and Putnam's attempts to deal with the problem of representation. Putnam argues that there is a direct form of perception, which he calls "natural realism," that escapes these problems, effectively waving aside the issues postmodernism raises. Rorty, for his part, is more sympathetic to postmodernism and suggests quite an opposite solution: "to give up on the idea that knowledge is an attempt to *represent* reality. Rather, we should view inquiry as a way of using reality. So the relation between our truth claims and the rest of the world is causal rather than representational."[6]

Dialogics is a philosophical and sociological tradition with a different origin that gets to much the same place as pragmatism, but in a distinctive way. While pragmatism is largely an American tradition—the British philosopher H. O. Mounce calls pragmatism "the distinctively American philosophy"—dialogics has mainly been developed elsewhere, principally through the almost entirely independent work of Mikhail Bakhtin in Russia, Paolo Freire in Brazil, and the Austrian-born Martin Buber.[7] Dialogics starts from the premise of the centrality of communication to human life. In Bakhtin's words, "To be means to communicate. Absolute death (non-being) is the state of being unheard, unrecognized."[8] Dialogics shares with pragmatism a focus on the everyday, with all its differences from person to person and situation to situation. Dialogics is, as Michael Gardiner has described, "a *practical* rationality, rooted in the concrete deed, and not detachable from specific situations and projected as some sort of speciously and decontextualized 'Truth.'"[9] As Paulo Freire put it, "There is no true word that is not at the same time a praxis. Thus, to speak a true word is to transform the world."[10]

Dialogics places equal emphasis on both difference and sameness, and their interrelations. As Bakhtin wrote, "I realize myself initially through others: from them I receive words, forms, and tonalities for the formation of my initial idea of myself. . . . Just as the body is formed initially in the mother's womb (body), a person's consciousness awakens wrapped in another's consciousness."[11]

Consequently, there is what Bakhtin called an "internal dialogism" to every word. "The word is born in a dialogue as a living rejoinder within it," he wrote.[12] Words are "territory shared," wrote his associate Valentin Volosinov.[13] We call words forth from each other, taking into account who the other is and what the other has said. We shape words for each other, as best we imagine each other to be. The words that I write at this moment are shaped by my best imagination of who will be reading them. Readers in this way shape what I am writing with them in mind. This process of taking the other into account is the basic phenomenological requirement of dialogics.[14] Thus, while it emphasizes difference and stresses the importance of difference in making what we communicate relevant for everyday life—while it is in this sense relativistic—dialogics is not solipsistic.[15]

Internal dialogism also leads to a different response to the problem of representation raised by postmodernism. Rorty argues that we must give up on representing reality, as this pursuit inevitably leads us to the false universalisms of the correspondence theory of truth. But dialogics says that representation is not correspondence. In dialogue, we represent in order to present an aspect of the world, to bring it into the present conversation, in a way that relates to others' experiences as it relates to our own. We don't present the world as it is in some sort of impossibly detached and objective way. We present it in a way that is relevant to the experiences and current situation of all parties in a dialogue, as best we can. As Freire notes, in so doing we immediately transform what that aspect of the world in fact is, by changing its consequences. And in so doing, we cultivate knowledge as we cultivate ourselves.

Dialogics also offers a theory of power, something pragmatism is largely silent about: This is the problem of monologue, of speaking without taking the other into account, enacted through the structures of conversation. All dialogue contains at least traces of monologue. Every word is a mini-monologue. There are also monologues at the largest scales of human conversation and knowledge cultivation, silences that are externally imposed. Just as problematic is that the everyday world is structured by our sociohistorical context. That context then comes to shape dialogue through a kind of hidden monologue that guides the representations we make and do not make, silences that are internally imposed.

Yet, just as there is no perfect dialogue, there is no perfect monologue. As Bakhtin noted, "every word is directed toward an *answer* and cannot escape the profound influence of the answering word that it anticipates."[16] Every monologue has to be addressed to an audience and the responses of that audience. Monologue, too, requires more than one person. Herein lie the seeds of critique. As monologue must make some connection to the everyday, practical life of its audience if there is to be communication at all, it must always leave itself open to the audience's response—even if that response is one of passive, quiet resistance.[17] Every answer we make to what another has said, even if the other had the most dialogical of motivations, is a response to its monologic dimensions. In formulating a response, we critique the other by relating the other's words back to our own practical situation.

A critical response, however, depends upon our trust that we in fact understand our own practical situation, and upon our sense of identification with that situation and its conversations. This is why phenomenological rupture and the dissolution of trust through the recognition of interests—monologic interests—is so crucial to changes in our cultivation of knowledge. We need to recognize monologue to critique it, which is no easy task, as we must recognize it from within our own context, the only context we have. The good news is that all responses we have in life depend upon at least a small measure of this recognizing, this re-knowing, of the monologic dimension of even the most dialogic word. To be means to communicate and to communicate means to critique—to offer our own difference to the world. We have this critical capacity, however effectively or ineffectively we may come to use it.

It's a never ending condition, and in this pragmatism and dialogics are of nearly one mind. Dewey called it *continuity*. "The principle of continuity of experience," he wrote, "means that every experience both takes up something from those which have gone before and modifies in some way the quality of those which come after."[18] Or, as he put it on another occasion, "a human society is always starting afresh. It is always in process of renewing, and it endures only because of renewal."[19] Bakhtin called it *unfinalizability*. In his words, "There is neither a first word nor a last word and there are no limits to the dialogic context."[20] Thankfully so.

CODA: *Sustaining Cultivation*

"You're missing something."

Dick Thompson and I were sitting at his kitchen table, a well-thumbed copy of the first draft of this book in a stack between us. Dick noted the flash of alarm that passed across my face.

"I like the book. Don't get me wrong. I like it a lot. But at least when you're talking about me, there's something you're missing."

"What's—" I started to ask.

"Get along but don't go along."

This is a phrase that Dick says he heard one day when cleaning a hog waterer on his farm, when no one else was around.[1] It's long been his motto, and he repeats it at almost every event he speaks at. I had clean forgotten about it, I suppose because I had never really underlined it in my mind. It seemed to me an interesting but quirky turn of phrase.

"Get along but don't go along. It's what your book's about."

He blazed those clear blue eyes at me. I met them for a moment, and then turned away. Then, suddenly, I got it. Finally. Dick saw his meaning land, and his face spread with a three-hundred-acre grin.

He was right. Get along but don't go along is Dick's way of saying, in just six words, something that has taken me nearer to a hundred thousand to say. The phrase neatly captures the tensions between modernism and postmodernism, difference and engagement, stability and change, that PFI farmers seek to resolve. Get along: Here Dick means both getting along with others and getting along in the world. He means to emphasize the importance of community and dialogue as well as the importance of achieving one's practical needs, and how these are interconnected.[2] Don't go along: Dick's point here is not that one should never do what others do, nor that one should, solipsistically, disregard what others have to say. Rather, his point is that one shouldn't do it just for the sake of getting along. In the terms I have been using in the book, just going along would lead to monologue and monologic power, to the suppression of difference and of dialogue's creative power. Dick's motto instead stresses the value of retaining and encouraging difference as *part* of getting along, as part of good relations and the achievement of practical outcomes. Falsely treating all our circumstances as the same leads to both alienation and irrelevance. Consensus is not necessary for getting along. Dialogue is, and for dialogue we need to retain a measure of difference if we are to find interest in what others have to say, and thus cultivate new knowledge as we cultivate their interest in and commitment to what we too have to say.

The view from Highway 71, north of Audubon, Iowa. 2001.

Which is just what PFI tries to cultivate. Their vision of sustainable agriculture is that it brings people together in dialogue—bouncing ideas off one another, testing propositions against the practical needs of growing food and feeding people in ways that protect the environment—keeps communities vital, and yields secure incomes. PFI does not offer its members the seductive comfort of final answers. Silver bullets are bullets just the same, no less hazardous for their gleam. PFI gives its members people to talk to and a means for comparing practical circumstances, ever adapting others' words to their own thoughts and offering their own words back for others to chew over, to reshape in light of their experiences, and to give back to others once again. It is not a process of adoption-diffusion. It is a process of *adaption*-diffusion.

The draw of monologue, of adoption-diffusion, of the Big Ag way should not be underestimated, though. Like the farmer at the wheel of a big green machine, monologue can give those who accept it a feeling of great power, even though the tractor is something that the farmer did not make and could not make, but merely switches on. It is a pleasure of power, not of control. That is to say, it is a borrowed power, an identity on loan. Although it is borrowed, in an uncertain world many feel they can hope for little more.

Moreover, most farmers find themselves trying to live on borrowed time with this borrowed power. So many acres, so little time. The speeding of the treadmill associated with the Big Ag way is itself a structure of monologue, encouraging farmers to seek the quick answers, the readily available answers, the silver bullets, even if one has a slightly uneasy feeling about the shooters of those bullets. No one can rely on their own experience for everything. We simply don't have time to reinvent the many wheels of our lives. We all have to rely on what others tell us. It's inescapable, unless your plan is to stay in bed for the rest of your life. But the problem gets worse and worse the faster the treadmill spins. And given the local intensification of this treadmill through the farmer's problem, farmers are increasingly less likely to look to their neighbors for answers, lest either party gain an advantage over the other thereby. Instead they turn to the monologic answers that the treadmill itself provides (only sending it spinning all the faster). The Big Ag phenomenology of doing agriculture thus becomes as well a phenomenology of knowing agriculture, of the cultivation of knowledge, of the cultivation of the ignorable, of the cultivation of the self, in this case a largely monologic self.

Cultivating oneself within the monologue of Big Ag does not ultimately ease the lives of most farmers, despite its lure of easy answers. The bankers of borrowed power and borrowed time occasionally call in the loans—not every year, and not on every farmer. But there are payments to be made here, even if they are not in the currency of money alone. This is a deeply disorienting experience for those who have little connection with other conversations of life and self.

To change metaphors, the structures of agriculture are a tough ride, and a lot of farmers get bucked off. It's a real shock when it happens—what I called in Chapter 6 a phenomenological rupture. When you're flying through the air, everything you know and do and are comes suddenly into question. Most farmers hit the dirt in a bit of a daze, get up, dust off, and see little else to do but to try to get back on the rampaging horse, if they can manage to get a foot in the dangling stirrup and with a mighty heave swing their way back up onto the saddle. And if they can't . . .

But while you're flying through the air, if the winds are right, you may hear, at just the right moment, of a safer way to land on the rodeo dirt, and of a whole different way to ride when you dust off and get back up. (Maybe you'll even hear of a different, gentler horse.) Not every farmer does. But when you're flying through the air, you listen to the voices in the crowd like you never have before. This moment of phenomenological rupture can actually be a kind of opportunity. So many of the sustainable farmers we interviewed described their decision to change practices as a sudden event. Given the connections between the structures of doing and the structures of knowing, and given the connections of these to the structures of self, there is so much that has to change that very often it needs to be all changed more or less at once—if it is to be changed at all.

PFI is one of those voices in the crowd, and it is a voice with a distinctive manner of speaking. It is a dialogic voice. It is a voice that tells you that you don't have to go along to get along. It is a voice that the farmers of PFI have come, not to adopt as their own, but to adapt to their own.

Farming isn't easy for the members of PFI either. There is still plenty of bucking with the horses they ride as well. Some PFI members eventually wind up leaving the rodeo of farming too. Despite all the environmental, economic, social, and possibly health benefits of what they do, the honest truth is that many of them fail. In fact, several of the PFI farmers that appeared in earlier chapters are no longer farming. Roger. Raelyne. Wendell and Terri. I don't know of a survey comparing the survival rates of sustainable and conventional farms, but my impression is that it's much the same for both, although possibly on average sustainable farms survive somewhat longer. Several PFI members I asked about this agreed. If there is a difference in survival rates, it is not so significant that, without a statistical survey, it jumps out at even those closely familiar with the matter.[3]

Sustainable farmers face several disadvantages in the farming rodeo. To begin with, they often start from a position of financial weakness. As we saw in Chapter 6, economic factors often act as the phenomenological shock that opens farmers up to the possibility of sustainable farming. Such farmers may find themselves just too banged up financially to regain a secure seat in the saddle, although sustainable practices may enable them to stay "in" farming a few more years. Also, as sustainable farmers tend to farm fewer acres and buy fewer outside inputs, such as machinery and chemicals, ironically they will be more likely to find themselves beholden to banks and merchants, rather than having the banks and merchants equally beholden to them. Because sustainable farmers spend less and borrow less, the balance of beholden-ness is more often tipped against them, making it more probable that the banks and merchants will call in their accounts.

Another disadvantage sustainable farmers face is with the structure of agricultural subsidies. The average Iowa farmer does quite nicely in this respect, with that annual subsidy check of $22,400 per farm between 1999 and 2001.[4] But averages disguise a lot. Some 72 percent of Iowa's farm subsidies go to just 20 percent of the farmers.[5] Smaller farmers usually get a much smaller check, and big farmers a bigger one. It might seem obvious that smaller farmers would receive less, but the subsidy system is often defended as the savior of small farms. Moreover, small farms typically receive lower subsidies per farm acre than big farms, particularly when the small farms produce something other than corn and soybeans, corn and soybeans, and corn and soybeans, the two main subsidized commodities in Iowa. Fewer acres and different products being two of their typical characteristics, sustainable farms tend to lose out substantially on farm subsidies.

Take Jon and Heather, for example, the small-scale beef farmers from Chapter 6. They have been getting about $6,000 in annual subsidies recently, less than a third of the average per-farm subsidy rate in Iowa.[6] But they have a 320-acre farm, which is near the Iowa average of 344 acres. If they received the Iowa per-acre average, their farm would have averaged about $20,837 in subsidies from 1999 to 2001. Jon and Heather didn't get that because they mostly grow grass and hay for their organic, free-range beef herd—grass and hay because that's essential for free-range production and because they want to limit erosion-prone and chemical-hungry crops like corn and soybeans. A few acres of their steepest land are enrolled in the Conservation Reserve Program, a government program for keeping the most erodible land out of row crops. They got about $1,000 in annual subsidies for that. And they do grow a few acres of corn, plus some wheat and sorghum, both unusual crops for Iowa. They got about $5,000 a year for that. But they missed out on an additional $14,837 a year of support that a more conventional operation of their size would have gotten.

Given that they favor big farms and not agronomic efficiency; given that they encourage overproduction and not soil conservation and environmental protection, in most circumstances; given that they total an awful lot of money devoted to at best dubious outcomes for the public good; given all this, many have long proposed the restructuring of farm subsidies. At the time of this writing, there is considerable hope among those concerned about sustainability that we may have just taken a large step in the right direction with the passage in 2002 (and the partial funding in 2003) of the Conservation Security Program. This hotly contested title in the 2002 Farm Bill would pay farmers for a whole series of green enhancements to their operations, from building soil-control structures to encouraging wildlife. It's a manifestation of what some observers of agriculture call "multifunctionality"—the idea that the purpose of agriculture should go beyond the single-minded one of ever-increasing food production. Rather than paying farmers to do only that, through our food expenditures and through the structure of subsidies, we should pay farmers to do some important things that are not directly included in the price of food—things like conserving soil and providing wildlife habitat. The CSP program would pay individual farms up to $45,000 a year to green their operations, based on a tiered system with three levels of enhancements. The CSP, many people hope, is the breakthrough in farm structure that sustainable farmers have been waiting for.[7]

But others are not so optimistic. The CSP's structure turns out to be very complex. Many of the small farmers who have looked into it have despaired of keeping up with the paperwork necessary to sign up for the program and to document compliance with it. Larger farms are more likely to be able to hire outside help to sort through its tangle of rules. Also, at this writing it is not clear that the program will receive anything like the level of funding that was originally envisioned for it. Although a farm could conceivably get as much as $45,000 a year in support, on a per-acre basis the program would pay no more than $22.25 an acre, with full implementation at the level of the third tier—considerably less than farmers have been receiving recently for leaving their land in corn. There is also concern about how green some of the practices the CSP will pay for really are, in part because the rules are still up in the air. Although it will probably on the whole encourage greener practices on the farms that are able to cope with its paperwork requirements, it is too soon to say if the CSP will significantly increase survival rates among sustainable farms.

Another factor that may depress sustainable farm survival rates comes, paradoxically, from a strength of PFI's get-along-but-don't-go-along approach. A farmer who has tried to seat his or her self on the tractor seat of the Big Ag monologue is also likely to be a farmer with a more unitary understanding of his or her self. The more dialogic self of a more dialogic knowledge cultivation, like PFI's, is a more multidimensional self, more accustomed to taking others into consideration and thus to envisioning what other selves might be like, including one's own self. While sustainable farmers may show no less commitment to being a farmer, to getting back on to that horse and to staying on it, they are also possibly more ready to consider other commitments when they find themselves slipping from the saddle once again. With a more open sense of who they are, they may be more ready to accept the outcomes of economic and other structural difficulties because they have come to see their senses of self as less dependent on those structures. They may be more ready to follow a path of self-reinvention, like one male PFI livestock and grain farmer who recently left farming to become a nurse.

But many of these points on the survival rates of sustainable farmers are at best informed conjecture. What we do know is that many PFI farms are prospering, perhaps as well as can be expected in such a high-risk, low-return endeavor as farming. We also know that other farmers are not exactly switching to sustainable practices with great alacrity. It is perhaps a steady trend, but it is still a slow one.

"I'm going to change the question a little bit about issues," said Donna, moving on to one of our standard questions. "What do you think are the most important issues facing farming today? The important issues facing rural families in America, even the world?"

"That's a tough one. I don't know where to start." Brad paused to collect his thoughts. It was one of those open-ended questions that is so broad one hardly knows what to say at first.

"One issue revolves around the polarization that's taking place," Brad said finally. "Where out in rural America, maybe for all of America, for that matter, the extremes have the loudest voice. Extremes on both sides. In agriculture, it's the farmers that are getting bigger all the time. It's the corporate aspects of farming that are getting bigger, especially with hogs now. And then the other side is still those of us that are choosing to farm much differently, and trying to sell our food more directly to the consumer. And so you got that group versus the corporate type of farmer."

Donna nodded and thought about saying something, but decided to stay quiet and let Brad's words tumble out more of their own accord. Meanwhile, Brad hesitated, perhaps to consider where he had gotten to, and perhaps to wonder if he felt right about calling his own side of farming as much of an "extreme" as corporate farming. In any event, he clearly was concerned about monologic interactions between the two sides, each "versus" the other, competing to be the "loudest voice."

"And then you got the rest," he continued. "Most of us are still caught in the middle yet, trying to figure out just where in the heck are we going."

"Right," said Donna supportively.

"It's just such a critical time, I think, for family farm agriculture. 'Cause I think we do have the opportunity to form networks and marketing groups and so forth, so that we can survive. But that's going to have to happen in the next five to ten years, or the opportunities will be gone. Then I think the other, the tremendous hurdle, is education of consumers and citizens in general about the value of farming the way we do."

In other words, unless the dialogue of practical agriculture widens well beyond its current confines, the cultivation of broad agricultural sustainability is unlikely to come to pass anytime soon. As Brad noted, part of that broadening has to take place among farmers and others in the farming community, such as university researchers, government officials, bankers, agricultural implement dealers, agricultural implement makers, seed suppliers, and commodity groups like the Iowa Corn Growers Association. Although Brad rightly worries that there is still much polarization here, much of the success of PFI has been exactly in its ability to invite the participation of others in the farming community into its conversations, into its knowledge cultivation, through the attractions of the group's dialogic approach. PFI has had the greatest impact on its local state university, Iowa State. It can be no accident that Iowa State in 1989 established the Leopold Center for Sustainable Agriculture, which is widely recognized as one of the nation's leading research and extension centers in sustainable agriculture. It can be no accident that in the fall of 2001 Iowa State University enrolled students in the nation's first graduate program in sustainable agriculture.[8] PFI has also made connections with many of Iowa's traditional agricultural organizations, such as the Iowa Farm Bureau, with which it jointly hosts a series of field days every year to farms that are implementing sustainable practices. At its 2003 annual meeting, PFI hosted a session, led by a rural banker, on the role of bankers in sustainable agriculture. PFI understands what I earlier called the "wonderful" quality of dialogue: that the inclusion of each additional voice in a dialogue encourages consideration of whomever else is missing, leading to the even greater widening of participation.

But Brad also pointed to another, equally important dimension of broadening the dialogue of practical agriculture: the "tremendous hurdle," as he put it, of involving those who eat and use what farmers produce in the discussion over what agriculture has become and what it could instead be.

To overcome that hurdle, to encourage the broader relevance of agriculture, we will have to consider the arguments for why agriculture is irrelevant. I think we can divide these arguments into three general sorts, what we might call the *emptying* of agriculture, the *swamping* of agriculture, and the *obsolescence* of agriculture. All three of these arguments, I believe, are misleading.

The emptying of agriculture is probably the most familiar of the three, and it goes like this. Ours is now an urban world. Some 75 percent of Americans live in urban areas, and 25 percent in rural areas, the reverse of how things stood a hundred years ago. True, worldwide, some 53 percent of the human population remains rural.[9] Each year, however, the rural percentage drops, as more and more of the human more and more find their fortunes where the real fortunes seem always to be made: in cities. And those who remain in rural areas, particularly in the wealthy countries, are increasingly apt to take up pursuits similar to those of their city cousins—factory work, retail sales, government services. Moreover, rural people in rich countries and poor increasingly commute to urban areas for work. Hardly any rural people actually still farm, at least in the rich countries. In the United States, farmers and farm workers are down to a couple percent of the workforce of the entire country.[10] Even in Iowa, where agriculture is a larger part of the economy than in any other state, farmers and farm workers account for only about 6 percent of the employed workforce.[11]

The swamping of agriculture is a related point, although the direction of the argument runs the other way. It suggests that as people have left farms for the city, the forces of the city have taken over the farms. There is nothing special about agriculture anymore. It is an industry like any other, and a business like any other, too. In the words of the rural sociologist William Friedland, "what is now called 'agriculture' has become mostly sets of industrial processes physically located in the open air rather than under a roof." Agriculture has been "transformed beyond recognition," Friedland says.[12] Increasingly, agriculture is a bad neighbor that local residents protest because of its pollution, just like a factory. And with the coming of "pharming," in which what is grown is not food but bioengineered medicines, agriculture is no longer "just like a factory"; it *is* a factory. There is no room here for any sentiment concerning community ties among rural residents, or between rural and urban residents. Agriculture is just a form of capitalism. Indeed, there is no reason even to call it agriculture anymore, suggests Friedland, as what we think that word means no longer exists.

Besides, as the agricultural economist Stephen C. Blank argues in *The End of Agriculture in the American Portfolio,* we don't really need agriculture anymore. Agriculture is obsolete. That's why there's no money in it, and that's why farmers are in such decline. Only a few consumers are willing to spend for their food what it would take to provide American farmers with a decent standard of living. "In the simplest terms, the production of food and other agricultural products will disappear from the United States because it will become unprofitable to tie up resources in farming and ranching," Blank argues. Although "many of them will not believe it at the time," he continues, many farmers "will, in a number of ways, be better off after they make the difficult decision to leave agriculture voluntarily." Good riddance to the high labor, high risk, high capital, and low profit of farming. "We need to strip away the romance and nostalgia surrounding agriculture and see it for what it is: a business," Blank writes. "We must learn to let go of farming and ranching."[13] We won't starve. Farmers in developing countries, with their lower wages and lighter environmental regulations, will be able to export plenty of cheap food to us, Blank argues. And as long as there is plenty of edible goo for us to microwave up whenever we require it, why should we care?[14]

Now, it must be said that there is some truth to these three arguments for the irrelevance of agriculture in the United States and other developed countries. More people do live in cities today, almost three-fourths of the developed world. The rural people that remain lead lives that are indeed much the same as those of urban folk, albeit with perhaps a bit more driving (or perhaps not, given the length of some urban commutes). We must admit that much of agriculture, if not most of it, hardly seems like farming anymore. To repeat the sociologist Paul

Lasley's phrase, agriculture is increasingly just Ag—ag business, ag chemicals, ag machinery, and perhaps just plain agony for some, given the stress, the struggle, the loss of economic control, the loss of community, the loss of environment, the loss of culture. Also, the United States has for many years been the world's largest food importer.[15] We are wealthy enough that we could probably easily import more, if need be.

For many, these trends are to be decried and resisted, lest we truly lose the real meaning of agriculture and the close interaction with communities natural and human that it affords and preserves. Indeed, this book might be read as such a decrying and resisting. And in a way it is. But in a way it isn't, for I believe that some of the concern to save agriculture is misplaced.[16] To decry and resist the "emptying," "swamping," or "obsolescence" of agriculture is, in some measure, to accept the terms by which these three visions of the end of the agriculture frame our understandings. And to accept the terms of the end-of-agriculture debate, whether pro or con, is to accept an unhelpful presumption: that agriculture is something that farmers do.

As Wendell Berry has written, "eating is an agricultural act."[17] To eat is to shape the contours of farmland as effectively as any tractor. What we eat is what we grow. How we eat is how we grow. Why we eat is why we grow, and there are many reasons for eating in addition to attending to the necessary, if generally pleasurable, sustenance of our individual bodies. The pleasures and necessities of eating, as we understand them, may be as much about the connections it makes with other places and other lives as about the filling of the self.

Among those connections are connections to the rural places that yield the bulk of what we eat, and to the people who plant the seeds, tend the plants, and harvest the crop. In this sense, there has been no emptying of agriculture. There is just as large a percentage of people involved in agriculture as there has always been: all of us. There may be fewer farmers, but there are no fewer agriculturalists. Indeed, given the increase in the human population, there are substantially more.

Nor, in this sense, can we say that agriculture has been swamped by the forces of the city. There is nothing nonagricultural about urban living. Without agriculture, there would be no urban living, and indeed little human living of any kind. Agriculture is indeed in many ways much different from what it once was, but that's largely because urban life has become part of it, not because urban life has become a source of agriculture's demise.

Nor is agriculture obsolete. If we were to switch to importing all our food, that would not make agriculture any less a part of the American economic portfolio. We would still be eating, paying for it, and shaping the lives of farmers and the land thereby. Those farmers and that land would just be that much further away, making our sense of connection to, and understanding of, eating as an agricultural act that much further away in our minds as well.

For while it is true, as long as we yet eat and live, that agriculture is not over, nor even close to it, it is also true that most of us no longer feel a part of its conversation.

In short, agriculture has largely become monologic. It speaks to us, not with us, and that goes for farmers as much as eaters. And it speaks to us with one logic: Cheap food is all we should ask of agriculture. We don't want to pay more than we have to for our food, of course. This much is true. But the single-minded focus on producing vast quantities of food in order to feed the world at low cost keeps us from seeing the full connectedness of the agricultural conversation, the connectedness both between people and between people and the earth.

Among those connections are the implications of the cheap-food monologue for farmers, in both rich countries and poor. In the United States, our agricultural subsidies, as we have seen, encourage ever-increasing production, which has the effect of driving down the prices farmers receive, which in turn leads them to try to stay on the treadmill and solve the farmer's problem by increasing their production even more. But although this process is often praised for keeping food costs down, it no longer has much impact on what eaters pay. Currently in the United States, about 19 cents of every dollar spent on food goes back to farmers—about $123 billion of the $661 billion American consumers spend on food each year.[18] Decreasing that percentage even further would have little effect on food prices, as it is now such a minor fraction of the food dollar.[19] But it would almost certainly result in fewer farmers on the treadmill.

In fact, we are currently awash in food in the United States, and in most other industrialized countries. The situation is so bad that, in order to prop up the prices farmers get, so that they don't all fold at once, we burn corn in our cars in the form of ethanol and we have an aggressive food export policy. In addition to being the world's biggest food importer, the United States is the world's biggest food exporter. Which is supposed to help feed the world. But as I discussed in the Overture and Chapter 1, most American agricultural products are too expensive for any but the relatively wealthy in other countries to purchase, the very people who are already eating pretty well. And in circumstances where American food—grain, primarily—does get to poor regions of the world, it tends to depress prices in those areas below what local farmers can make a living on, putting them out of work and making them even poorer. "Feeding the world" with American grain is consequently often just a pretty slogan for commodity dumping.

Despite being awash in cheap food at home, there is nonetheless widespread hunger in the United States. The U.S. Department of Agriculture categorizes nearly 35 million Americans, including 13 million children, as "food insecure"—approximately 12.5 percent of the nation in 2002. Approximately 9.4 million live in households that the USDA classifies as "hungry"—about 3 percent of the nation.[20] This hunger persists despite the fact that Americans spend less of their income on food—about 11 percent—than practically any other industrialized nation. But those are average figures. Poor people need to eat as much as rich people do, and although they cut costs and spend quite a bit less, they still spend a much higher proportion of their income on food. People making between $5,000 and $10,000 a year spend about 33 percent of their after-tax income on food.[21] Also, poor people in the United States typically face higher food costs, as supermarkets today typically are located in suburban areas, away from where most poor people live.[22] The solution to hunger is not to increase food production in the United States. Rather, it lies in reducing inequality here and abroad, in increasing food production within poor countries, and in protecting the environmental base of food production.

Also, most of us don't need more food. In the United States there are perhaps greater health problems from overeating than from undereating, as the recent national discussion of obesity has underscored. We need better food. Given that we don't pay much for food to begin with; given that farmers don't get much of what we do pay; given that we currently spend our subsidies mainly to support high production, with relatively less attention to quality production—given all that, better food should be well within the nation's capabilities, without a significant increase in food costs, if any at all.

And we've been working on it, perhaps most notably through the recent growth in the production of organic food. Once found mainly in the form of tired-looking, expensive veggies in hippie co-ops, organic food is now big business. Some 39 percent of U.S. consumers report that they use organic products at least occasionally.[23] However, organic products are typically significantly more expensive and generally do not yet deliver on the promise of higher quality food with only a small price increase, or even a price decrease. Nor is organic production necessarily helping very many small farms to stay afloat, as organic increasingly becomes bought up into what the writer Michael Pollan has aptly termed the "industrial-organic complex."[24]

The point is, as Patricia Allen and Martin Kovach note, organic "is simply not enough."[25] Organic agriculture in itself does little to reconnect eaters with the agricultural conversation. In fact, it can be seen as largely a confirmation of their disconnection. What I mean here is that organic food retailing commonly trades on consumers' growing sense of unease with their disconnection from the agricultural conversation, and their consequent lack of trust in what they eat. Organic certification is narrowly based on production criteria. Buying organic food likely does support environmental protection and probably animal welfare, too, but in itself it has little to do with economic justice and bringing people back into the agricultural conversation. The focus of the organic label and its associated inspection system is on individual consumer health in an untrusting world—on what the sociologist Melanie Dupuis has termed "not in my body" sensibilities—more than reconnecting eaters with a dialogic, practical agriculture.[26]

We have become used to seeing agriculture as a realm out there in the beyond that we have left behind, physically and culturally. This beyond and behind notion of agriculture envisions it as a space, something that we enter when we leave, or perhaps escape, the sprawling city and the artifice and ambition we sense there. Agriculture now feels distant, something that we rarely see and that little impinges on our daily activities. Agriculture feels like a product line—shall I have the Cortlands or the Fujis or the Granny Smiths?—not a relationship to other humans and the earth. That feeling of the product line is part of the feeling of distance, for disconnection is distance.

So why not bring agriculture back home, to everyone's homes? The pleasures of this connection are what Jack Kloppenburg, John Hendrickson, and Steve Stevenson mean when they speak of "foodshed" thinking. The increasingly globalized food system of today encourages a sense of placeless food. Sure, we might know that in North America oranges come from Florida in the winter and from Brazil in the summer, and it may even say so on the signs in the supermarket produce aisle. But here place is used mainly as a sort of brand. We get little sense of the lives of those who raised, sorted, and shipped these oranges. We get little sense of the environmental implications of their methods. They're just oranges, some cheaper, some more expensive; some sweeter, some a bit more dried out. But thinking of the food we eat as flows in a foodshed, like flows of water in a watershed, gives direction to the movement of food, and thus an origin to food in specific places and in the specific lives lived there. Foodshed thinking makes all food homemade, for it connects us to the home places of what we eat.

Foodshed thinking also teaches where we ourselves are. "The foodshed is a continuous reminder that we are standing in a particular place; not anywhere, but here," write Kloppenburg, Hendrickson, and Stevenson.[27] It leads to what Thomas Lyson has called "civic agriculture"—a "locally-based agricultural and food production system that is tightly linked to a community's social and economic development," in Lyson's words.[28] And it is leading to it with a will. Communities across the world, especially in those places where food had become the most place-less, the most monological, are returning to local foods, and the dialogic connections to people and environment that is their sweetest taste. In the United States, we have seen an enormous growth of farmers' markets, farm stands, pick-your-own, community-supported agriculture (CSA) projects, community gardens, community farms, community kitchens, institutional buying of local products, and small-scale food processing. For example, between 1994 and 2002 the number of farmers' markets around the country rose from 1,755 to more than 3,100, a 77 percent increase.[29] There are now more than a thousand CSA, or subscription, farms in the United States, with more than a hundred thousand member households.[30] Although they are sometimes threatened by development pressures, the growth of community gardens and community farms has been bringing agriculture right back into the city, making the eaters the growers.[31] People are coming to savor the taste of place and the enjoyments of eating locally, eating seasonally, and eating in ways that support local farmers and local communities.

The point of local agriculture is not that North Americans need to give up tea, coffee, and bananas. The point is that, in a food system in which what we eat may come from thousands of miles away and where the typical food item is handled thirty-three times from field to supermarket shelf, there is abundant room to cut way back on the "middlemen" so as to give a greater share of the food dollar to farmers, while giving eaters healthier, tastier, more environmentally friendly, community-supporting, and quite possibly even cheaper food.[32] And everyone gains a sense of connection, of dialogue between grower and eater.

PFI has been among those groups that have in recent years worked hard to provide that connection and dialogue. Through its Field to Family project, PFI has worked to combine issues of equity and sustainability with local eating and local sourcing of food. PFI instituted a "healthy food voucher" program that enabled low-income families to participate in the CSA project the group helps coordinate. PFI has organized local institutions to use local sources of food, such as the campus conference center at Iowa State, which now provides an "all-Iowa meal" option for organizations that use the facility. PFI regularly conducts cooking classes and nutrition classes so that eaters can regain the skills of healthy and efficient home food preparation. PFI started up a farmers' market in Ames, the city where Iowa State is located and where PFI's own offices are. PFI started an annual youth summer camp centered on sustainable agriculture and sustainable food, and began a gardening and nutrition program with the Ames Boys and Girls Club.

But perhaps the greatest effort that PFI has made to connect farmers and eaters came in 2002, when its members voted to allow nonfarmers to become full voting members of the group. The group still asks people to state on their membership application whether they gain "a significant part" of their income "directly from farming in Iowa." And only those who check this box can serve on the group's board. So it is still at heart a farmers' organization. But anyone can join and vote, and half of the group's roughly seven hundred members are eaters, not farmers, who have taken up this invitation to engage in the dialogue of what a practical agriculture could look like—this invitation to find a place in the wordshed of the foodshed.

The broad sense of connection underlying foodshed thinking allows us to dispense with another unhelpful presumption about agriculture: that its purpose is only to grow food (and fiber and, I suppose, medicine now). We do want agriculture to grow food for us (and fiber and, perhaps, medicine). But growing food is only one dimension of what I would argue is the purpose of agriculture: *cultivation*—the care and tending of creation, human and nonhuman, social and ecological. Here I mean cultivation in a way different from but related to how I have used it earlier in this book. Here I mean it not as the relationship between who we are and what we know, as the culture of identity and the identity of culture. I mean cultivation as the culture of the earth, as the husbandry and wifery of life, as farming for us all.

Cultivation, then, is a task not only for those people we have long regarded as the agriculturalists. It is a task not only for farmers. It is a task for everyone. I worry, though, that the image of the farm tends to guide our thinking back to a sense of agriculture as the beyond we left behind. There couldn't be a farm in a city, right? A city couldn't itself be a farm, could it? There is something jarring to the pattern of our imagination here.

Thus I suggest, along with Harriet Friedmann, that we imagine the dominant metaphor of agriculture as that of the *garden*, the garden writ large, for a garden is something that I believe we are more used to understanding as potentially everywhere, in city and countryside alike.[33] We are more used to identifying ourselves as potentially all gardeners than as potentially all farmers. Moreover, there is an intimacy and care associated with gardening, an intimacy and care that Big Ag has, to a large extent, driven out of our image of farming. And indeed most farms today do not feel remotely like gardens. Nor do most of our towns and cities, our schools and neighborhoods, our workplaces and our public life, at least

Virginia Moser, PFI *member, at the* CSA *farm she operates with her husband, Marion Moser, 1999.*

not much of the time. These too are increasingly a form of Big Ag, farming us all, not farming for us all.

So let us put the culture back in agriculture of all forms and in all places. Let cultivation of what Friedmann has termed the "gardens of Gaia" become our understanding of what the rural is and what farming should be—of what we sometimes succeed in and so often fail to attain in actual social and ecological life in both country and city.[34] To speak of agriculture, then, in this ideal sense, is to consider the degree to which a state of mind and action brings out, or fails to bring out, the gardener's capacity of care for creation.

Which brings us back to dialogue and sustainability. It is something of a romance, I freely confess, to consider agriculture as cultivation in this largest sense, as care of the earth. But it is a romance that allows us to appreciate how the current structures of production so overwhelm our better intentions for the cultivation of the garden. It is a romance with material consequences. It is a practical romance. But only if we all engage in the conversation of agriculture.

For we probably do not all agree about what it means to care for the earth and its ecological and social creation. And that is fine. More than that: It is great—as long as we engage our differences and take responsibility for response ability, for difference is creativity's own unfinalizable wellspring. By coming to imagine agriculture as here and now, wherever we are, and not beyond and behind, we ask for and welcome that engagement. Eaters, growers, purveyors, teachers, builders, doctors, lawyers, bus drivers, factory workers, scientists, even realtors and regulators: We are all potentially agriculturalists in the sense of the gardening of creation. It depends upon our state of mind and action, upon how we contribute to this task that is, thankfully, larger than any of us.

This, I believe, is the practical message of Practical Farmers of Iowa: that cultivation in the agricultural sense depends upon an approach to cultivation in the social sense that embraces the creativity of difference, openness, and the unfinalizable. A dialogic, practical agriculture requires that we all consider agriculture a central feature of everyone's lives, worthy of everyone's care and careful attention. A dialogic, practical agriculture requires that we sustain the broadest possible conversation about agriculture. For what Iowa's practical farmers are really trying to farm is democracy, the democracy of the people's good earth. The same is true of practical farmers everywhere. They are all guided by a common insight: that farming for us all, in every field of human endeavor, is only possible when there is farming by us all.

ACKNOWLEDGMENTS

As I sit down to write these acknowledgments, the autumnal equinox fast approaches and the harvest is on in the Midwest. It's a time for finishing up, for gathering together, for putting things away, for reflecting on the season past. As day and night near equality, it's a time for assessing the balance of things. And it's a time for giving thanks.

Harvest and the equinox thus seems to me a very good time to write the acknowledgments for a book long in cultivation, perhaps especially for a book about farming. When the crop is full ripe (or as ripe as it's going to get), it needs to be gathered together. And so it is with the crop of this book. It is time to finish up. It is time to reflect on the season past and to give thanks for those who have contributed so much to its fruits.

A common myth—there can be no other word for it—is that the real farmer farms alone, toiling single-handedly against the elements, and that every farm has only one farmer. He (in the myth it almost always is a he) is the farmer. The others are farm workers, farmhands, farm wives, farm kids. The members of Practical Farmers of Iowa and other sustainable agriculture groups, as this book describes, are struggling to overcome this singular understanding of the authorship of a farm, as part of their efforts to create a more dialogic, and thus more practical, agriculture. We need also to struggle against the parallel myth of the single authorship of a book. No author truly works alone. No book truly has just one author.

This is particularly true of a book such as this one. Central to whatever fruit this book can be said to offer are the participatory methods with which the underlying research was conducted. Without Donna Bauer and Sue Jarnagin of Practical Farmers of Iowa and their on-the-ground insights, and without Greg Peter's master's thesis work, this book would have been not only less informed, less intelligent, less relevant, less valid and reliable a witness of the conditions of farming today—to the extent it has any of these good qualities—it would have been a whole lot less fun. To them I owe the biggest thanks.

Long I wrestled with how to present Greg's, Donna's, and Sue's contributions to the book. At first I tried writing the narrative in the "we" voice throughout. Yet that proved stylistically awkward, especially for presenting the firsthand accounts that make up much of the text. The book does use the "we" voice here and there, but in the end we all agreed that it would be most honest, direct, and inviting to use the "I" voice for the most part. After all, I wrote up the text, and it must be said that, for better or worse, my voice dominates the interpretations and analyses the book presents.

Next I tried writing the "intermezzo" sections of the book as dialogues among the four of us, in the classic style of the ancient Greeks. I felt that dialogue form would give Sue, Greg, and Donna their own first-person voices in the text. But although it took me a year to see it, which held up publication of the book, I eventually came to agree with what most readers of early drafts of the manuscript said. However worthy the intent may have been, the dialogues came across as distracting, cutesy, and self-indulgent.

These narrative failures make me uncomfortable, but I'll just have to live with them. As well, I would have loved it if our work on this book more closely matched the kind of participatory equinox that was my ideal. But it didn't. Lives and their commitments and constraints do not always, or even generally, afford such evenness of participation. The word "with" puts it well, though, and not just in the publisher's sense. I am not *the* author. Donna, Sue, and Greg are authors of this book with me. I was not alone in the work on this book, and I am very, very grateful to them for that.

There were other participants, too, many, many others, and therefore many other authors of this book with me. The next mention must be of the dozens of Iowa farm families who opened their homes, farms, and lives to us. Their words wrote most of this book. Unfortunately, the need to maintain the confidentiality with which we conducted the research prevents me from mentioning most of them by name in the text or in these acknowledgments. (A few Iowa farmers are mentioned by name in the text, but only in descriptions of public events, or in a couple of instances with their specific permission.) But my gratitude for their contributions is none the less heartfelt for that.

Two farmers I can and must mention by name here for their steadfast support and guidance are Vic Madsen and Dick Thompson, both former presidents of Practical Farmers of Iowa. Vic was president when I first proposed writing this book and was enthusiastic about it from the start, helping me with contacts among PFI members and in the broader Iowa farming community and providing me with insight after insight in his warm, quiet way. Dick was PFI's first president and remains its best-known member—a tireless advocate of sustainability, on-farm research, dialogue, and getting along but not going along. He too provided me, as he has so many, with insight after insight about farming conditions in Iowa, and gave a draft of this manuscript the most helpful of critical readings.

And then there's Rick Exner, PFI's first and longest-serving employee. He's not a farmer in the conventional sense of the word, but he is a farmer. He has spent years cultivating the dialogues that have made PFI's successes possible. Talk about years: My dialogues with him about this book began in 1994. An enormous proportion of what I know about farming in Iowa I owe to those dialogues.

This book simply could not have happened without his interest and confidence in it. Rick also gave a draft of the manuscript a detailed reading, saving me from many an embarrassing error and enriching the argument in many places.

Pennsylvania State University Press and the Rural Studies Series of the Rural Sociological Society have also been steadfast in their interest and confidence in this book—as was much needed, given the time it has taken me to complete it. (Talk about years again.) My editor at Penn State, Peter Potter, and my editor at the Rural Studies Series, Leif Jensen, were paragons of the gentle critic, cheering section, and, when needed, fire-breathing dragon. I tried their patience many a time, I think. Fortunately, they had a lot to try.

One of the services that Peter and Leif performed for this book was picking two fabulous reviewers for the manuscript. Happily, they have both identified themselves to me so I can thank them publicly for helping me grapple with a host of nettlesome problems, not least the proper format for the infamous intermezzos, the quite unsatisfactory first draft of the final chapter, and my tendency to overwrite. I may not have been able to completely resolve these problems (and others they identified), but this would have been a much poorer book without their interventions. So thanks indeed to Melanie Dupuis and Carolyn Sachs, reviewers extraordinaire.

And then there's Mitch Duneier. Everyone should have a colleague like Mitch, and I count myself very fortunate to be someone who does. Mitch was an incredible advocate for this book, both with potential publishers and even with me, when my own confidence and interest in it flagged. He was also very helpful in getting me through some of my hang-ups about the modernism-postmodernism debate (although quite possibly he doesn't remember this). And when reactions came back from various readers of the manuscript, Mitch was a great sounding board. (It is an irony in both our lives that I moved from Ames to Madison just as Mitch was leaving Madison. Oh well. I guess I'll forgive him.)

Another thing Mitch did was convince me, in part through the wonders of his own books, of the many values of visual ethnography. It was uncommon good fortune, then, to discover among the membership of PFI a most excellent photographer, Helen D. Gunderson, whose black-and-white images so grace this book. Helen's love of Iowa farming and rural life, their promise and challenges, shows in every dot of every halftone here.

I was also fortunate enough over the years to get feedback on the emerging manuscript from Allison Brooks, Jerry DeWitt, Jess Gilbert, the students in Jess's class on the sociology of agriculture, Gary Guthrie, Maynard Haskins, Gary Huber, Robert Karp, Matt Liebman, the students in the class Matt and I taught together at Iowa State, and Steve Weiss. I give thanks for them all.

There was important institutional support for this book, too. In 1995 the Sustainable Agriculture Research and Education (SARE) program of the U.S. Department of Agriculture awarded me the grant that supported Donna's, Sue's, and Greg's work on the project. My own work on it was mainly supported by Iowa State University and its Department of Sociology, where I served from 1993 to 2002 as a member of the faculty of the College of Agriculture. In the spring semester of 2001, Iowa State awarded me a sabbatical to the Centre for Study of Agriculture, Food and Environment (CSAFE) at the Department of Anthropology at the University of Otago in New Zealand, where I completed about half the writing of the book, in restful escape from an office phone (but not from e-mail). CSAFE's director, Hugh Campbell, was an astonishingly good host in an astonishingly good country. Back at Iowa State, Will Goudy and Bob Schaffer both supported this book in the countless little, and sometimes big, ways that one's department chair can do. Also at Iowa State, a number of people struggled with transcribing nearly a hundred hours of tapes from our field work. They include Rachel Burlingame, Leslie Daub, Margie Hanson, Beth Hoff, Lori Merritt, and Virginia Wadsley (in addition to the hours put in by Greg, Sue, me, and especially Donna). Thanks, thanks, and thanks.

I suppose it's not exactly an institution, at least in the formal sense of the word, but I can't help also offering my thanks to the Pretty Good Band, the all-PFI group I often played with at PFI gatherings over the years: Rick Exner on guitar, Mary Sand on fiddle, Joe Lynch on concertina, Lonna Nachtigal on hammered dulcimer, and me on banjo and mandolin. What am I thanking them for that has to do with this book? A heck of a lot of fun with great folks. That has a lot to do with this book.

Finally, let me give thanks to my family, Diane Bell Mayerfeld and our children Sam and Eleanor. Diane has long been my best critic, and Sam and Eleanor are fast joining her there. Life with them is a continuous seminar, a bit chaotic at times but nonetheless a tumble of joyous and loving debate and dialogue. They teach me so much. They put up with me so much. The smallest of them is playing with my toes as I write this sentence, yanking and pulling and asking me if she's squeezing too hard yet. I get the point. Time to get back to the family seminar. Hit the save button.

—Madison, September 2003

NOTES

Overture

1. I believe Dick is paraphrasing a line from Wendell Berry here, although I have not been able to find the source.

2. Iowa's average acreage in corn and soybeans over the period from 1995 through 1999 was 22.4 million acres (12.3 million corn, 10.1 million soybeans). Sands and Holden (2000).

3. I give the subsidy figures here per acre of farmland, not per acre of corn and soybeans, the two main subsidized crops in Iowa, in order to better reflect the impact of subsidies on Iowa agriculture in general. Over the period from 1995 to 1999, direct government payments to Iowa farms averaged $1.006 billion per year. Iowa averaged 33 million acres in farms during this period, which works out to a per-acre subsidy of $30.48 per year. However, from 1999 to 2001 Iowa averaged $2.115 billion in annual subsidies, or $64.08 an acre. If these dramatically higher payments continue—and the 2002 farm bill, just passed as of this writing, would increase them a further 70 percent—the trends toward industrialization in Iowa's agriculture will accelerate. Note that typically all but a few percent of the cash value of Iowa crops comes from corn and soybeans. If one were to calculate the per-acre subsidy on the basis of the 22.4 million acres in corn and soybeans that Iowa averaged from 1995 to 1999, the figure would be $47.01 per acre; using the same acreage for 1999 to 2001 (and there was little change), the subsidy is $96.14 per acre. (In fact, the subsidies in Iowa have historically been based on crops, mainly corn, although FAIR—the "Freedom to Farm" Act of 1996—made this connection less direct.) By contrast, during the period 1995 to 2001, the estimated cost of crop production for corn exceeded the average market price farmers received in five out of seven years (averaging together the cost of crop production for corn following corn and corn following soybeans); and in one of the two years in which prices exceeded costs, 1996, the margin was only 3 cents an acre. The situation was somewhat better in soybeans, with estimated costs exceeding prices in four of seven years. Thus, without subsidies, corn-and-soybean farming is a break-even proposition at best. Liebman (2001) argues that even with subsidies corn–and-bean farming is a break-even endeavor, unless farmers can find a way to beat the estimated costs of production. All my numbers are from Sands and Holden (1998, 2000), Environmental Working Group (2002), and Duffy and Smith (2002).

4. Under the "Freedom to Farm" policy of the 1996 Farm Bill, with its sly acronym of FAIR (for the Federal Agricultural Industry Reform Act), these subsidies were slated to progressively disappear by 2003. But lawmakers on both sides of the aisle were shortly afterward calling Freedom to Farm a failure, not least because, bizarrely, the amount paid in subsidies actually went up after the law was enacted, as Congress rushed in with supplementary subsidies with every new farm crisis. The level of subsidies in 1999 was actually the highest ever, and 2000 wasn't far from it. At the time of this writing, the new farm bill, the Farm Security and Rural Investment Act of

2002, has been recently enacted, and it is projected to increase subsidies by some 70 percent over the unprecedented figures of the past few years.

5. Median household income in the United States in 1999 was $40,816 for all households; $43,342 for households with two people; $51,190 for households with three people; and $59,768 for households with four people. The figure for households with one person was substantially lower, $21,083, pulling the median figure down. Few farm households, however, are this small. Figures are from U.S. Census Bureau (1999).

6. This figure of 92,500 for the total number of farms in Iowa in 2002 comes from the Iowa office of the U.S. Department of Agriculture's National Agricultural Statistics Service (NASS), which calculates an annual figure (Iowa Office of the National Agricultural Statistics Service, 2003). The U.S. Census of Agriculture, which uses slightly different methods and definitions, reports the number of Iowa farms as 90,792, based on the 1997 Agricultural Census, the most recent one at this writing. NASS calculated that Iowa had 98,000 farms in 1997, some 7,000 more than the U.S. Census Bureau reported for that year. But both sources agree that there are roughly 90,000 to 100,000 farms in Iowa, and that the number is steadily declining, year after year.

7. Dick's experiments in night planting date from 1996 to 1997, if I recall correctly. If you visited the Thompson farm today, you'd probably also see a small Norwegian plow, which they use occasionally to prevent stratification of nutrients in their soil. Plowing is anathema to many who are concerned with the same issues of conservation that the Thompsons are. But the Thompsons believe that, properly and carefully used, a plow still has its place on a well-managed and sustainable farm and that it does not necessarily promote erosion, at least on their mostly level farm. Recently the Thompsons have also been experimenting with a flex-harrow; modifications of their Buffalo cultivator; the use of engine exhaust to control rats in their corn bin; adding hydrogen peroxide to their water supply for disease control; weeding by the phase of the moon; and much more. They are constant tinkerers.

8. Davidson ([1990] 1996).

9. The breakthrough publication was the 1989 report by the National Research Council of the National Academy of Sciences, *Alternative Agriculture* (National Research Council, 1989), and the "scientists' review" of the report a year later (Jordan et al. 1990). For recent overviews, see Altieri (1995), Beeman and Pritchard (2001), Bird, Bultena, and Gardner (1995), Brookfield (2001), Collinson (2000), and Gliessman (1998).

10. A technical historical note on flame cultivation: Some in the sustainable agriculture movement see it as a new technique, but in fact the use of flame cultivation dates back to the 1930s and has long been used in cotton production in the South. Similarly, there is much new attention to direct-marketing farm products, but it too is an idea with old roots. The historic depth of ideas, however, should not detract from our appreciation of the innovation with which they are being brought to bear on current conditions.

11. The precise figure of Iowa farms that raise livestock, as of the 1997 agricultural census, was 57 percent (National Agricultural Statistics Service, 1999, table

2). However, the number is dropping fast. Many thousands of hog farmers gave up during the steep price dip during the winter of 1998–99, and thus the percentage of Iowa farms with livestock now is probably below 50 percent.

12. An official at the Iowa Department of Agriculture and Land Stewardship (IDALS) informed me that IDALS is the originator of this number but now entirely repudiates it as empirically meaningless—which it is.

13. For details, see Chapter 1, note 11.

14. For an introduction to the growing sustainable agriculture literature, see Altieri (1995), Beeman and Pritchard (2001), Berry (1977), Francis (1990), Gleissman (1998), Jackson (1985), Jackson and Jackson (2002).

15. This is what Robert Bogdan and Steven J. Taylor (1990) call "optimistic research."

16. For this sort of argument, see Blank (1998), Hart (1991), and Tweeten (2003).

17. I do not mean that farmers are free to be inefficient producers, though. My point is that most farmers do strive for efficient production, because of its economic advantages, and most attain a similar level of efficiency, adopting the same or similar practices as their neighbors. There is striking homogeneity in the agricultural practices of Iowa grain farmers. Among these farmers, the relatively slight differences in their efficiency are of considerably less value than the level of their subsidies in predicting whether they will stay "in." It is a common conclusion in agricultural circles that if a farm goes under it must have been because of inefficient production, but the main evidence of that inefficiency is usually that the farm went under. I take up the problem of this tautological reasoning in more detail in Chapter 1.

18. Environmental Working Group (2000).

19. Throughout the rest of this section, among other things, I am engaging in a theoretical dialogue with the literature on "local knowledge" and "farmer knowledge," and most especially the work of Jeff Bentley, Neva Hassanein, and Jack Kloppenburg. For an introduction to this literature, see Bentley, Rodriguez, and Gonzalez (1994), Chambers (1994), Collinson (2000), Hassanein (1999), Hassanein and Kloppenburg (1995), and Kloppenburg (1991). For a more explicit critique of Hassanein (1999), see the first Intermezzo.

20. Knowledge and power "directly imply one another," the French philosopher Michel Foucault famously observed ([1975] 1979, 27). Thus Foucault liked to speak not of knowledge but of "power-knowledge." Yet it was Foucault who made this observation, and it would be regarded as a considerable infringement on human trust not to mention that. The phrase "power-knowledge" is a cultivar of knowledge that we locate socially by attaching it to the name Foucault—or, at the very least, to the community of scholars, and more specifically to the community of postmodernist scholars.

21. For an introduction to participatory research and the history of its development, see Chambers (1994), Gaventa (1993), Kemmis and McTaggart (2000), Stoecker and Bonacich (1992), and Bell (1998b).

22. A long line of scholars have made this point, from Georg Simmel ([1908] 1971) at the turn of the nineteenth

century to, more recently, Patricia Hill Collins (1986).

23. Leopold ([1949] 1966), 210.

Intermezzo

1. "Writing it down" is what the ethnographer does, Clifford Geertz (1973) once observed, but this is true for every act of scholarship, as all scholarship is ethnography of one sort or another.

2. Foucault ([1975] 1979, 27).

3. Chambers (1994).

4. The full quotation from Hassanein (1999, 10) is "farmers and other advocates of sustainable agriculture have created an alternative knowledge system that functions primarily outside of the dominant institutions of agricultural research and extension."

5. For examples of calls for a bottom-up response, see Brecher (1994), Johnson (1992), Whitley (1992).

6. Hassanein (1999, 189–90).

7. For a poetic response to such a view, see Bell (2002).

8. Cf. Musambira (2000) on postmodernism and cross-cultural communication, and Fenton (2000) and Press (2000) on postmodernism and feminist communication theory.

9. I take the phrase "practical postmodernism" from Gardiner and Bell (1998, 7). For an exegesis of the related concept of "postmodern pragmatism," see Depew and Hollinger ([1995] 1999). The work of Richard Rorty is often so described, but Rorty (1999) has expressed considerable doubt about the helpfulness of the term "postmodern," and always insists on placing the term in quotation marks.

10. Beck, Giddens, and Lash (1994).

11. Beck, Giddens, Lash, and other critics usually describe the problem of postmodernism as relativism, but the problem is really that of solipsism. The embrace of "reflexivity" by Beck, Giddens, and Lash shows their own welcome of at least a degree of relativism.

12. Indeed, the kinds of responses to modernism that the term "postmodern" seeks to describe are as old as modernism itself. Postmodernism, in varying degrees, has been with us as long as we have had modernism in its own varying degrees.

13. Beck, Giddens, and Lash are, in fact, well aware that reflexiveness is not reflexive as yet. Beck, for example, devotes much of his work to describing the struggles of what he terms the "subpolitics" of questioning science, technology, and economy. But if this is so, they should be more up front about "reflexive modernization" being more optimistic than empirical. Even more complexly, Gilbert (2003) points out that at least the rural aspects of the progressive movement in the United States during the 1930s were highly participatory, a kind of "low modernism" that was at least as reflexive as the supposed period of "reflexive modernization" we are currently entering. Reflexivity, like postmodernism, is nothing new.

14. I also feel that there is an unfortunate pun in the word "reflexive." It sounds more like reflex—a nonreflective response—than reflection.

Chapter 1

1. In order to provide anonymity, I have altered certain details of the J and D café.

2. See Otto and Lawrence (2002).

3. For a review, see Thu and Durrenberger (1998).

4. On the time pressures of contemporary family life, see Hochschild (1997) and Schor (1999). On downsizing, see

Dudley (2000). On the conflict between home and work see Hochschild (1997) and Nippert-Eng (1996). On struggles to accommodate difference, see Stein (2001). The decline of community is an old theme in American social science, recently given new force by the widespread interest in social capital, most notably through the work of Putnam (2000).

5. Malone (2002).

6. There have actually been several land grant proclamations since the original one in 1862, most notably the 1890 act that established the "black land grants" across much of the South, and the 1994 executive order that incorporated the tribal colleges on Native American reservations into the land grant system.

7. In many states, however, the traditional College of Agriculture has been renamed something along the lines of College of the Environment and Life Sciences, as at the University of Rhode Island, or College of Life Sciences and Agriculture, as at the University of New Hampshire, or College of Agriculture and Life Sciences, as at about ten universities.

8. Extension programs also traditionally draw on colleges of home economics, now often renamed "College of Family and Consumer Sciences" or "College of Human Ecology." At the University of Wisconsin–Madison, though, extension is set up as a separate college. Also, in most universities, faculty from any college can participate in extension, although usually most come from colleges of agriculture and home economics, in their contemporary renamed forms.

9. Blank (1998) to the contrary notwithstanding.

10. According to the Bureau of Labor Statistics (n.d.), there were 6.3 million construction workers in the United States in 1999. The number fluctuates widely, though, and has ranged from 4.5 million in 1992 to 6.9 million in 2001.

11. For number of farmers, I round off here the 1999 figure of 2,048,400 given by the National Agricultural Statistics Service (NASS) for "self-employed and unpaid workers" in agriculture, which is the standard measure used. In that year, the United States also averaged 929,000 hired farm workers. There is also a widely fluctuating number of "service workers" in agriculture, ranging from 157,000 to 319,000 in 1999. In that same year, NASS tabulated 2,192,070 farms in the United States, slightly more than the number of farmers. It would be tempting to see this slightly higher number as indicating the influence of corporate agriculture, but the definitions behind both these numbers are so fraught with problems that it is best to be content that they are roughly comparable. See National Agricultural Statistics Service (2002) and Bureau of Labor Statistics (2001).

12. USDA (1998a, table 1).

13. USDA (1998a).

14. USDA (1998a, table 1), and Goudy et al. (1996, 7).

15. Otto et al. (1997).

16. From calculations I made, based on Otto et al. (1997).

17. I am aware that I am running against the grain of much economic analysis here, whether of a conventional sort based on the notions of "value added" and indirect impacts or whether more ecologically oriented, such as the "picnic principle" of Jacobs (2000)—the idea that the most efficient and productive systems are ones in which every crumb goes to something, even if only

ants. But if every analysis of the economic benefit of particular economic sectors were added together, I suspect we would find our economy accounted for many times over. There is a seductive tendency to boosterism in this kind of economic research, making all such studies suspect.

18. Bryson (1989, 3).

19. Paul Lasley is a colleague of mine from Iowa State University in the Department of Sociology. I thank him for this observation.

20. Williams (1973).

21. Bell (1992, 1994, 2004).

22. The phrase "treadmill of production" comes from the classic work of Alan Schnaiberg (1980). Similar perspectives can be found in Willard Cochrane (1958) and James O'Connor (1973), in keeping with Marx's identification of the "crisis of over-production" in the structure of capitalist enterprise.

23. By the phrase "farmer's problem," I mean in part to refer to the related version of the "prisoner's dilemma" that McEvoy ([1986] 1990) called the "fisherman's problem" and in part to refer to Cochrane's (1965) discussion of the "farm problem."

24. Although the crisis is perpetual, several useful studies pay special attention to the social consequences of the 1980s period, including Barlett (1993), Dudley (2000), Harper (2001), Lasley et al. (1995), Lobao (1990), and Davidson ([1990] 1996).

25. Unfortunately, government agricultural programs have rarely grappled with this problem, usually preferring to work on issues of production and demand.

26. Fifty-four percent, to be precise (USDA, 1998a).

27. On the "treadmill of consumption," see my own work, Bell (1998), as well as the work of Schor (1999) and Durning (1992).

28. See Chavas and Aliber (1993) and Jaforullah and Whiteman (1999). For work that finds big dairy farms more efficient, see Richards and Scott (2000) and Cocchi, Bravo-Ureta, and Cooke (1998).

29. Logsdon (1994, 120).

30. There is, of course, an enormous literature on this point. For a recent critical review based on a rereading of the work of Adam Smith, see Korten (1999).

31. Dr. John Lawrence, Iowa State University livestock economist, personal communication.

32. The agricultural economist Mike Duffy tells me that the correlation between monopolization and price cycles has not, historically, always been so clear, however.

33. A congressional inquiry concluded that market access is being limited in this and other ways, and Senator Charles Grassley of Iowa led an unsuccessful effort to include correcting legislation in the 2002 farm bill.

34. Reported in Anthan (2000). See also Eisenberg (n.d.).

35. Hendrickson and Heffernan (2002) estimate that in 2001 59 percent of the pork-packing industry was controlled by just four firms, and 75 percent by just six, with Smithfield Foods leading the way. See also Heffernan, Hendrickson, and Gronski (1999).

36. Barboza (1999).

37. I thank Mike Duffy for this observation. See the Coda for a more extensive discussion of its implications.

38. There is an extensive economic literature on this topic. Even sociologists, who ought to know better, have often made this argument. See, for example, Rogers (1995, 269), and see the sociological literature on the "adoption" and "diffusion" of innovations, and my critique of this literature at the end of Chapter 6 and again in the Coda.

39. See Hightower (1973).

40. This raises the question of why a fuss was made over the 1980s farm crisis. The answer is that it wasn't so much a farm crisis as a banking crisis, as I argue in Bell (1999).

41. We are also losing the independent banks once common in small rural towns, as these are themselves bought up by bigger firms with little sense of local commitment, and little tolerance for accepting a lower rate of return so that the parents of the banker's children's schoolmates can stay in business. I thank Rick Exner for this observation.

42. USDA (1998a).

43. Jackson-Smith (1999).

44. Duffy often gives talks at farmers' meetings. I have heard him use this line on several of these occasions.

45. Environmental Working Group (2002) figures, divided by the 96,000 farms the U.S. Agricultural Statistics Service recorded for Iowa in 1999 (see Chapter 1, note 6). Anthan (2001) reports considerably higher subsidy figures, but the Environmental Working Group is more reliable.

46. Environmental Working Group (2002) figures, divided by Iowa Office of the National Agricultural Statistics Service estimate of 96,000 farms in Iowa in 1999 (Sands and Holden, 2000).

47. Environmental Working Group (2000). Note that farm payments under the Freedom to Farm Act accounted for about 75 percent of the total federal farm payments to Iowa. An additional 25 percent came from loan deficiency and market loss assistance payments, but no analyst that I have read argues that this additional 25 percent was allocated so as to equalize payment levels.

48. Anthan (2001). Duffy was speaking about the programs under the 1996 farm bill, but the 2002 farm bill actually has made the tendency toward inequality in farm subsidies worse, "nailing the little guy" even more.

49. This study, conducted by Sparks Companies, Inc., of Memphis, Tennessee, was presented to Congress in February 2001. See Anthan (2001).

Chapter 2

1. There is a vast literature arguing that community no longer depends upon space. This literature dates from well before the rise of the term "globalization." The classic reference is Webber (1963). For recent reviews, see Noller (2000) and Gieryn (2000).

2. I do not here mean to imply that there is, in fact, any greater degree of community in rural areas. There is an old debate in social science about whether there is more community in rural life than in urban life, and the empirical evidence suggests, rather unequivocally, that there is not. However, when you talk to them about it, people living in rural areas do typically place a large stress on the importance of local community, in part for reasons of identity, as I argue in the case study reported in Bell (1992). Here I suggest some material reasons why, in addition to issues

of identity, local community is so often emphasized in rural conversations.

3. See the discussion in Chapter 1.

4. On this point, see Bell (1992).

5. I thank Rick Exner, Farming Systems Coordinator of Practical Farmers of Iowa, for this information.

6. Lay, Haussmann, and Daniels (n.d., but based on experiments conducted in 1998.)

7. Brewster et al. (n.d., but based on a 1998 study) found that hoops required significantly more labor. But another study by Practical Farmers of Iowa, in conjunction with Iowa State University agricultural economist Mike Duffy, found that hoops require only slightly more labor, Rick Exner, Farming Systems Coordinator of Practical Farmers of Iowa, informs me.

8. This conversation took place in 1996. Iowa experienced a slight rise in its average wage during the second half of the 1990s and into the early 2000s, like most of the country.

9. Iowa Quality Pigs is a pseudonym, but many hog confinement operations have similar names.

10. Becker (2002).

Chapter 3

1. My reference to an "American dilemma" is to that classic study of race relations in the Unites States, Myrdal (1944). On tension between home and work, see Hochschild (1997) and Nippert-Eng (1996).

2. See Hochschild (1997).

3. Nippert-Eng (1996).

4. Goffman ([1974] 1986).

5. Schegloff (1996) provides a good account of the theory of conversation analysis. For a classic empirical application, see Duneier and Molotch (1999).

6. USDA (1997).

7. The practice goes by several names, including "rotational grazing," "rotational intensive grazing," the "New Zealand system," and "Voisin controlled grazing management" (see Hassanein 1999, 50). The latter name is in testament to André Voisin, a member of the French Academy of Agriculture who conducted extensive research on the system in the 1950s (Voisin, [1959] 1988). It is often described in the social science literature as a classic example of "farmer's knowledge," given that it was farmers who mainly introduced it into the United States in the late 1980s and early 1990s. In New Zealand, however, agricultural scientists promoted the practice among farmers, based on agricultural experiment station research from the 1950s. And Voisin, although he owned a farm, was a trained biochemist. But Voisin and the New Zealand researchers themselves drew on principles noted by farmers for centuries. The question of whether grazing represents "farmer's knowledge" (or what is often called "local knowledge") or "scientific knowledge" is an example of the process of knowledge cultivation discussed in Chapters 5 through 8.

8. The best study on this topic, although it is now a bit dated, remains Hochschild ([1989] 1997).

Intermezzo

1. This is the analytic problem methodologists call "validity"—the extent to which a method of research accurately represents the object under study. Methodologists commonly go on to distinguish validity from "reliability"—the extent to which another research project using the same methods with the same research object would get the same

results. It is a well-known aphorism among those who teach social science methods that quantitative research tends to be highly reliable but of questionable validity. Qualitative research, for its part, tends to be highly valid but of questionable reliability.

2. Duneier (1999).

Chapter 4

1. Crop production figures from Duffy and Smith (2002).

2. See Liebman (2001) for an argument that grain farming in Iowa is fundamentally a money-losing business, even after adding crop subsidies in on top of the regular sale prices for grain.

3. On the hypermasculinity of advertisements and promotions to farmers, see Brandth (1995). Recently, however, pesticide companies have been switching to more "sensitive" names for their products, like Axiom, Clarity, Galaxy, Liberty, and Synchrony.

4. On the importance of granting everyone the capacity, and thus the power, to secure their legitimate interests, see Sen (1992) and, in a way, both Walzer (1983) and Rawls (1972). On the importance of the farm as a source of autonomy in a class-structured world, see Mooney (1988).

5. On the importance of thinking of masculinity in the plural, see Connell (1995). On the significance of this plural conception for understanding the rural masculine, see Campbell (2000), Campbell and Bell (2000), Kimmel and Ferber (2000), Law and Dolan (1999), Leipins (2000), Little and Jones (2000), Meares (1997), Peter et al. (2000), and Woodward (2000).

6. This was long true of rural researchers as well, and still is, albeit to a lesser extent, thanks to works such as Feldman and Welsh (1995), Flora (1985), Meares (1997), Sachs (1996), Tickamyer and Bokemeier (1988), Valdivia and Gilles (2001), and Whatmore (1991).

7. The 1997 agricultural census records the percentage of female Iowa "farm operators" as 5.1 percent, up from 3.9 percent in the 1992 census (USDA 1997). Based on our observations, 5.1 percent is almost certainly too high for how "farmer" is conventionally understood. The figure probably reflects the growth in widows who operate farms with contract labor as much as it does women taking up farming on their own. But this figure is almost certainly way too low as well, if we define "farmer" in a nonpatriarchal fashion, since women make significant contributions to the farm operation on almost all Iowa farms.

8. There is an extensive literature by now on personality differences between women and men, and the terms "relational" and "rational" are often used to describe women and men, respectively, following the work of Carol Gilligan ([1982] 1993). One can easily make too much of these differences, however, as Cynthia Fuchs Epstein (1988) observed some years ago. We also need to be wary of a ready tendency to essentialize them, as in popular books of the *Men Are from Mars, Women Are from Venus* sort. We are talking cultural habits here, not biological straitjackets, as de Beauvoir ([1949] 1989), Connell (1995), Haraway (1991), Lorber (1994), Smith (1987), Zita (1998), and many others have cogently argued to our largely deaf ears.

9. There is an extensive literature on this point. For one of the original works, see Boserup ([1970] 1989).

10. I depart here somewhat from the Weberian convention in sociological theory that sees class as separate from status. I prefer to consider class as a status and economic power as one of the metrics we use to identify different forms of status. As I note in Bell (2003, 76), "we typically remark upon gender, race, ethnicity, abilities and disabilities, and [other forms of status] when we see them too aligning with economic power, or when we are startled to see an economic contradiction of these dimensions of social honor, as with the poverty of the elderly."

11. There is a considerable literature now on suburbia as an attempt to capture a lost rural idyll of the clean and green. For examples, see Bormann, Balmori, and Geballe (1993), Jackson (1985), Penning-Rowsell and Lowenthal (1986), and Teyssot (1999). However, we found ourselves wondering at moments like these if the reverse is not also true now, considering the spread of suburban values more generally among farmers and rural people.

12. On the "privacy" of farming, especially as a product of capitalist structures and ideologies, see Pile (1990).

13. At least not consciously: Sonya Salamon, in a well-known study, found that English and German farmers in Illinois have different attitudes toward land and land transfer across generations in a family, although they may not experience their attitudes as ethnic (Salamon, 1985, 1992).

14. Van den Berghe ([1967] 1978).

15. Delgado and Stefancic (1998), Mittelberg and Waters (1992), and Omi and Winant ([1986] 1994).

16. Bell and Griffin (2002).

17. This use of humor is what the Russian social philosopher Mikhail Bakhtin referred to as "uncrowning" ([1965] 1984). When one explicitly accepts "uncrowning," status differences seen as potentially hierarchical can become less important. For a further example of this from previous fieldwork of my own, see Bell (1994, 59–61).

Chapter 5

1. At least when I interviewed his family.

2. Many accounts of culture see it as a passive constraint on human imagination, while some others imagine it as an epiphenomenon of other processes. I, however, see culture as an active site of creativity, or at least potentially as this (Bell 1998b). It is in part to underline the activeness of the social relations of knowledge that I describe it as the "cultivation of knowledge" and not the "culture of knowledge," finding that the word cultivation more readily conjures up an active image, especially given the leaden-weight image that the word "culture" has acquired in many quarters.

3. The similarity of the two words is not accidental. The interrelationship between trust and truth is both social and etymological. For more on both topics, see Carolan and Bell (2003).

4. Davidson ([1990] 1996, 102) found much the same when he visited a basement meeting of a patriots' group, the Iowa Society for Educated Citizens. Davidson's observations of that meeting are very similar to my own.

5. See the history section of the American Bar Association's web site, accessed Nov. 10, 2003, at www.abanet.org/media/overview/phistory.html.

6. Davidson ([1990] 1996).

7. Dudley (2000, 164).

8. Davidson ([1990] 1996, 123).

9. This same focus on learning the "real truth" about things is reflected in the name of the group that Davidson observed, the Iowa Society for Educated Citizens.

10. For a subtle critique of the concept of hegemony, from the left, see Scott (1985, 1990). Scott argues that conventional hegemony theories vastly underestimate people's critical powers and often mistake covert resistance for compliance.

11. Although he didn't use the phrase, I'm pretty sure from other things Frankie said that the audience was supposed to understand "bankers" to be "Jewish bankers." Davidson ([1990] 1996, 103) observed much more explicit references to a supposed conspiracy by "Jewish bankers" in the Patriots meeting he attended.

12. I take the term "unknowledge" from the work of Sedgwick (1990), although her use of it is somewhat different. Working out of a Foucaultian cultivation, her interest is in the way power forces some knowledges into the closet, such as knowledge of homosexuality. That interests me too, but I seek throughout this chapter to balance the relationship between power and knowledge with an equal focus on identity, recovering the human agents of power and knowledge from the functionalist social vacuum of discourse theory. For more on this scholarly debate, see the discussion later on in this chapter.

13. James ([1896] 1927, 219). However, note here the individualism in James's account of the ignorable; the process of cultivating the ignorable is predominantly, if not inescapably, social. This is a more general problem with classical pragmatist theory, which I address in the fourth Intermezzo.

14. For example, here is the full thunder-and-lightning passage in Hosea that contains Frankie's favorite Bible quotation, albeit in a different translation: "My people are destroyed for lack of knowledge: because thou hast rejected knowledge, I will also reject thee, that thou shalt be no priest to me: seeing thou hast forgotten the law of thy God, I will also forget thy children."

15. Duffy (1999). Probably the number is significantly lower now.

16. There may also have been a class dimension to Gene's arguments—Gene has only a high school diploma, and I would guess that the animal rights position in part represents for him the ideas of the university crowd. But he didn't raise this explicitly or hint at it, as far as I could tell.

17. Foucault defined "power/knowledge" thus: "There is no power relation without the correlative constitution of a field of knowledge, nor any knowledge that does not presuppose and constitute at the same time power relations" ([1975] 1979, 27). Perhaps the most controversial feature of this influential formulation is Foucault's insistence that power/knowledge is the agentless product of discourse—that power/knowledge, through discourse, creates agents, and not the reverse. There is no position outside of power, he argued. "It seems to me that power *is* 'always already there,'" replied Foucault in response to a question in a famous interview, "that one is never 'outside' it, that there are no 'margins' for those who break with the system to gambol in" (1980, 141). Or, as he put it another time, "power is not an institution, a structure, or a certain force with which certain people are endowed; it is the name given to a complex strategic relation in a given

society" (Gordon, 1980, 236, giving his own translation of a line from Foucault). In other words, people don't have power (or power/knowledge); power (or power/knowledge) has people. The self here is banished as a false hope of the Enlightenment. This agentless power and powerless self eventually proved untenable even to Foucault, and in his last years he set about investigating how the self is constituted by discourse, and concluded that the self has "techniques" for its own self-constitution—that it is indeed able to gambol at times (Foucault, 1985, 1986, 2000). "For, if it is true that at the heart of power relations and as a permanent condition of their existence there is an insubordination and a certain essential obstinacy on the part of the principles of freedom, then there is no relationship of power without the means of escape or possible flight," he admitted in 1982, making theoretical room for politics (2000, 346). But in the few years left to him, he was not able to bring to unity this active self with its techniques and this discursive self without them, as even a sympathetic commentator like Strozier agrees (2002). The vessel of the Foucaultian self remained largely empty. It is with an eye to recovering a fuller, more dialogic, view of agency and the self that I introduce knowledge cultivation, the naming of knowledge, the social relations of knowledge, and the awkward neologism "power/knowledge/identity."

Intermezzo

1. Although I believe such a view is fading, it is still common to hear sociologists describe sociology as the study of social structure, and to regard cultural research as less important and less rooted in reality.

2. I offer here a general account of how sociologists have conceptualized culture. For a more detailed discussion, see Bell (1998b).

3. For a detailed critique of the view of culture as an epiphenomenon, see Bell (1998b).

4. My source here is Giddens (n.d.), his personal web site, as opposed to the earlier definition given in his famous book on the subject (Giddens 1984).

5. I refer here in particular to Bourdieu's concepts of "habitus" and cultural "fields." For an introduction to his work, see Bourdieu and Wacquant (1992).

6. Archer (1996).

Chapter 6

1. This passage demonstrates well the meaning that would have been lost had I chosen a less contextualized style of narrative. Without "novelistic" writing, Dale's "right, yeah" would be a mere transcript and would carry little of the meaning he intended.

2. Goffman ([1974] 1986).

3. Readers who think they hear an echo of Marx ([1852] 1972) in this line are right.

4. Thompson, Thompson, and Thompson (2001).

5. See Bell (1994, 1998a).

6. Gordon Bultena, Eric Hoiberg, and Michael M. Bell, unpublished data. Owing to changes in the researchers' circumstances, these data have never been fully analyzed, and I am not in a position to report on them in any detail.

7. Ibid., for the data on political affiliation and ethnicity. For the other points, see Bird, Bultena, and Gardner (1995). In our Iowa survey we were particularly surprised by the finding concerning ethnicity. A well-

known study by Sonya Salamon (1985, 1992) found that farmers of English heritage and farmers of German heritage had considerably different orientations toward land, with the German farmers more often regarding land as a patrimony entrusted to them to maintain for future generations in the family. We speculated that such an attitude might influence a farmer's concern for environmental issues and thus for sustainability. But we found no association.

8. Bird, Bultena, and Gardner (1995, 69). But note that in Montana, which has a significantly different agroecology, sustainable farms are far larger than conventional ones.

9. Ibid., 156.

10. This is actually one place where I find myself in agreement with the "diffusion of innovations" model of Rogers (1995), who also says age in itself has little to do with technological change.

11. See Chapter 3.

12. Dick Thompson and the others involved remember these meetings with great fondness, I later found out.

13. For more on dialogic unpredictability, see the concept of "diction" in Bell (2003).

14. This made me consider how many farmers in Iowa have, by now, heard the story of the College of Agriculture professor who couldn't tell soybeans from button weed.

15. Duffy and Smith (2002).

16. A number of states, including Iowa, Nebraska, and Ohio, have recently passed new regulations making it easier to compost dead stock on the farm.

17. I thank Rick Exner for bringing this saying to my attention.

18. There's enough scope for interpretive variation here that to be precise one should not be precise.

19. See Bell (1994, 1998a).

20. Much of the basic research in "adoption-diffusion" was conducted by sociologists at Iowa State University and the University of Wisconsin–Madison, the two departments where I have spent my professional academic career. Valente and Rogers (1995) trace the origin of the adoption-diffusion research tradition to the work of Bryce Ryan and Neal C. Gross, both rural sociologists at Iowa State University in the 1940s, and their pioneering paper (Ryan and Gross 1943). Joe Bohlen (who used to inhabit my former office at Iowa State) and George Beal carried this tradition on into the 1950s and 1960s, writing papers, including one with their now-famous student, Everett Rogers (Beal, Rogers, and Bohlen, 1957; Bohlen, 1967), as well as several influential reports (Beal and Bohlen, 1957; Beal and Rogers, 1960). In a widely read overview of the sociology of agriculture, Buttel, Larson, and Gillespie (1990) cite adoption-diffusion as the most important area of scholarship from the 1950s to the 1970s. Rogers (1995) claims that by the mid-1990s more than four thousand scholarly works had been published on the subject, which is an extraordinary total.

21. Rogers (1995, 266) confesses that "'laggard' might sound like a bad name," but then goes right on using it. He also suggests (1995, 269–74) that the characteristics of "early adopters" include greater "intelligence," being more of a "cosmopolite," being "less dogmatic," and having "greater rationality" and "greater empathy"—attributes that seem as loaded as "laggard."

22. Although he does not directly engage the adoption-diffusion tradition, the classic critique along these lines is the case Kloppenburg (1991) makes for the significance of "local knowledge," later elaborated in Flora (1992), Hassanein and Kloppenburg (1995), Roling and Wagenmakers (1998), Roling and Jiggins (1998), and Hassanein (1999). Other critics have complained of (1) a "pro-innovation bias" in adoption-diffusion research; (2) a tendency to blame individuals for failure to adopt, the rural sociological equivalent of "blaming the victim" that ignores the structural factors of social life; (3) the potential that adoption-diffusion research promotes social inequality by encouraging the targeting of potential early adopters in the promotion of a new technology; (4) a behaviorist and social-psychological bias associated with all of the previous shortcomings; (5) the inapplicability of the model to the Third World; (6) the inapplicability of the model in the First World, particularly in situations in which the profit motive does not come first, as in conservation technologies; (7) a total blindness with regard to issues of gender; (8) little attention to how learning takes place, particularly from a constructivist point of view; (9) a weak theoretical core; and (10) methodological problems associated with the mainly survey-based research used by researchers to study an inherently time-based process. See Buttel, Larson, and Gillespie (1990), Fleigel (1993), and Roling and Jiggins (1998) for summaries of most of these. Rogers (1995) accepts many of the critiques but then waves them aside. Others researchers read the implications quite differently and agree with Whyte (1985) that adoption-diffusion research has been a "failure." Although some adoption-diffusion research continues in rural sociology (for example, Padel 2001), for many researchers the tradition is something of an embarrassment for the profession and is steadfastly ignored and little discussed, except in jokes in the hallway about "laggards" and other adoption-diffusion lingo. Outside of sociology, however, adoption-diffusion remains extraordinarily popular, particularly in those fields not noted for their concerns about sociological issues such as inequality and blaming the victim.

Chapter 7

1. Duffy and Ernst (1999) and Duffy (2001) found that the economic "returns to management" of non-GMO soybeans were roughly equal to GMO soybeans—in fact, the returns to non-GMO soybeans were, if anything, a bit better. Evidence is also now in that "round-up ready" GMO crops are promoting the selection of herbicide-resistant weeds in farmers' fields (Pollack 2003).

2. Nor is it Foucault's panopticon (Foucault 1979). But it could well be argued that it seems more and more like it everyday.

3. With apologies to the late Douglas Adams.

4. Spontaneity, although it sounds similar, comes from the Latin *sponte*, of one's own accord, not *spondere*. However, the similarity of the words is, I think, a happy coincidence. On "response ability," see also the related concept of "answerability" in the work of Bakhtin, especially Bakhtin (1993). Answerability refers to the ethical commitment to respond, whereas by response ability I refer to the social conditions that promote that commitment and its practice.

5. For an introduction to the Bakhtinian approach to dialogue and the concept of unfinalizability, see Bakhtin ([1965] 1986; 1984a; 1984b) and Gardiner and Bell (1998). Bakhtin (1984b, 166) offers this widely quoted description of unfinalizability: "Nothing conclusive has yet taken place in the world, the ultimate word of the world and about the world has not yet been spoken, the world is open and free, everything is still in the future and will always be in the future."

6. The PFI site is http://www.pfi.iastate.edu.

7. Exner and Thompson (1998).

8. I make these statements based on the results of the group's own surveys of its members, as reported to me by Rick Exner.

9. This advertisement appeared in *Iowa Farmer Today*, NW edition, Dec. 1, 2001, p. 9.

10. For a time, Jerry DeWitt was even acting head of SARE. It is important to mention here that the bulk of the research for this book was conducted with funding from a grant from this program, awarded some years before this interview and before DeWitt became its acting head.

11. Figures from the office of the president, Iowa State University.

12. The classic works on the role of business interests in research in land grant universities are Hightower (1973), Hightower and DeMarco ([1976] 1986), Busch and Lacy (1983, 1986), and Buttel and Busch (1988). For a more recent overview, see Wolf and Zilberman (2001).

13. This pot of money came from a fine that Exxon Corporation paid to every state for overcharging for its oil, and was to be used for energy efficiency projects. A lot of innovative work across the United States was funded by these monies, while they lasted. PFI also received early on a generous grant from Mrs. Jean Wallace Douglas, one of the main inheritors of the Pioneer Hybrid fortune and one of sustainable agriculture's biggest financial supporters.

Chapter 8

1. I present here something of an amalgam of several Iowa small towns, so as to preserve the anonymity of the group members.

2. For our original discussion of dialogic masculinity, see Peter et al. (2000).

3. On gender as a practice, see Lorber (1994) and West and Zimmerman (1987).

4. Lyson (2000) refers to these new agricultural practices as "civic agriculture," highlighting the important public dimension of the relations they establish.

5. These are three of the main suppliers of organic vegetable seeds in the United States.

6. Wells and Gradwell (2001).

7. I draw these examples of how globalization may be increasing food risk from a recent popular review, Ackerman (2002). For more general critiques of fast food and the globalization of the food supply, see Bonanno (1994), Bové and Dufour (2001), McMichael (2000), and Schlosser (2002).

8. These estimates come from the Centers for Disease Control (Ackerman, 2002). Note that even without globalization of the food supply, there would be substantial incidence of food-borne illness for other reasons.

9. Freire ([1970] 1999).

10. The brand name is "Niman Ranch," http://www.nimanranch.com. Zinkand

(2001) describes the founding of the Iowa-based Niman Ranch Pork Company, a fifty-fifty partnership with Niman Ranch.

Intermezzo

1. Hildebrand (2003).

2. Peirce (1878).

3. James used this phrase throughout many of his writings, for example, James ([1925] 1953, 165).

4. Dewey (1928, 198–99).

5. Dewey (1939).

6. Rorty (1999, 33).

7. There has still been little scholarly effort made to engage these three visions of dialogics, however.

8. Bakhtin (1984b, 287).

9. Gardiner (2000, 53).

10. Freire ([1970] 1999, 68).

11. Bakhtin (1986, 138).

12. Bakhtin (1981, 279).

13. Volosinov ([1929] 1973, 86).

14. Bell (1998b).

15. For this reason I do not join in the common practice of criticizing some postmodernism for being relativistic. The problem is solipsism, not relativism. Difference is great, as long as we can also bring it together with other differences, through dialogue.

16. Bakhtin (1981, 280).

17. Scott (1990).

18. Dewey (1988, 13:18–19).

19. Dewey (1928, 429).

20. Bakhtin (1986, 170). See also the quotation in note 5 to Chapter 7.

Coda

1. Dick describes this experience as follows in Thompson, Thompson, and Thompson (2002, 1–3): "Several years ago while cleaning out a hog waterer, Dick heard a voice that said, 'Get along but don't go along.' There was no other person around at the time. This concept is what we are supposed to do. We don't have to convict or convince anybody, just share when asked. This makes the yoke easier and the burden lighter. This policy has left the door open to go to many land grant universities in the United States and overseas during the last few years."

2. This is also in keeping with what I have elsewhere termed the "dialogue of solidarities" (Bell 1998c).

3. Three members I asked said they thought survival rates of sustainable and conventional farms were about the same. A couple thought the rate was slightly higher for sustainable farms.

4. See Chapter 1, note 45.

5. Environmental Working Group (2002).

6. I looked up their subsidy on the Internet, courtesy of the Environmental Working Group, www.ewg.org, which has usefully posted the subsidy figures for every farmer in the United States.

7. See, for example, the web site of the Minnesota Project, a nonprofit Minnesota environmental group, www.mnproject.org/csp/.

8. I make this observation with some caution, as I was myself involved in establishing this program. Thus readers should take into consideration that I may have some personal interest in promoting its significance.

9. Figure for 1999 (United Nations Population Fund, 1999).

10. See Chapter 1, note 11, for details.

11. Based on dividing the number of farms in Iowa in 2002 (92,500, accord-

ing to the Iowa Office of the National Agricultural Statistics Service, 2003) by the size of the employed labor force in the state (1,560,300, according to Iowa Workforce Development, 2003), one gets a higher figure than the slightly less than 5 percent I earlier (Chapter 1, note 16) derived from Otto, Swenson, and Immerman (1997), to which I have added a percent or so for farm workers.

12. Friedland (2002, 352, 368).

13. Blank (1998, 1, 3, 193, 195).

14. I'm adapting a line here from the Iowa State University agricultural economist Mike Duffy (personal communication), who certainly is no fan of food as edible goo.

15. It is also the world's largest food exporter, a point that Blank does not consider closely.

16. For examples of the Jeffersonian view that I wish to sidestep, see Comstock (1986).

17. Berry (1990, 145).

18. Figures for 2000, from USDA Economic Research Service (2003).

19. I thank Mike Duffy, who has been heard from several times in this book, for this observation.

20. Nord, Andrews, and Carlson (2003).

21. Blisard (2001, table 19).

22. Kaufman et al. (1997).

23. Organic Trade Association (2003), citing a Natural Marketing Institute 2002 study.

24. Pollan (2001).

25. Allen and Kovach (2000).

26. Dupuis (2000).

27. Kloppenburg, Hendrickson, and Stevenson (1996, 41).

28. Lyson (2000, 42).

29. Lyson (2000, 45), and U.S. Department of Agriculture (2002).

30. Several web sites, including several USDA web sites, attribute this figure to a 1999 USDA study, but I was not able to locate the original source.

31. In perhaps the most dramatic example of the development of urban agriculture, some neighborhoods in Cuban cities now raise some 30 percent of their food locally. See Funes et al. (2002).

32. The figure of thirty-three times is from Kahn and McAlister (1997).

33. Friedmann (2003).

34. Ibid.

REFERENCES

Ackerman, Jennifer. 2002. "Food: How Safe? How Altered?" *National Geographic* 201 (5): 2–51.

Albrecht, Don E., and Steve H. Murdock. 2001. "Rural Environments and Agriculture." In *Handbook of Environmental Sociology,* ed. Riley E. Dunlap and William Michelson, 192–221. Westport, Conn.: Greenwood Press.

Allen, Patricia, and Martin Kovach. 2000. "The Capitalist Composition of Organic: The Potential of Markets in Fulfilling the Promise of Organic Agriculture." *Agriculture and Human Values* 17: 221–232.

Altieri, Miguel A. 1995. *Agroecology: The Science of Sustainable Agriculture.* 2d ed. Boulder: Westview Press; London: IT Publications.

Anthan, George. 2000. "Agriculture Secretary Fears Farmers Face 'Feudal' Future." *Des Moines Register,* March 26. Accessed May 19, 2002, at http://desmoinesregister.com/extras/country/feudal.html3.

———. 2001. "Bailouts Favor Big Farmers, Congress Told." *Des Moines Register,* Feb. 4. Accessed Feb. 8, 2001, at http://DesMoinesRegister.com/news/stories/ c4789013/13700684.html.

Archer, Margaret. 1996. *Culture and Agency: The Place of Culture in Social Theory.* Cambridge: Cambridge University Press.

Bakhtin, Mikhail. 1981. *The Dialogic Imagination: Four Essays.* Austin: University of Texas Press.

———. [1965] 1984a. *Rabelais and His World.* Bloomington: Indiana University Press.

———. 1984b. *Problems of Dostoevsky's Poetics.* Ed. and trans. Caryl Emerson, with an introduction by Wayne C. Booth. Minneapolis: University of Minnesota Press.

———. 1986. *Speech Genres and Other Late Essays.* Trans. Vern W. McGee. Minneapolis: University of Minnesota Press.

———. 1993. *Toward a Philosophy of the Act.* Austin: University of Texas Press.

Barlett, Peggy. 1993. *American Dreams, Rural Realities: Family Farms in Crisis.* Chapel Hill: University of North Carolina Press.

Barboza, David. 1999. "The Great Pork Price Gap: Hog Prices Have Plummeted; Why Haven't Store Prices?" *New York Times,* Jan. 7. Accessed Jan. 13, 2003 at http://oll.temple.edu/economics/econ52/news-clips/supply-demand/hog-prices.html.

Beal, George M., and Joe M. Bohlen. 1957. *The Diffusion Process.* Special Report 18. Ames: Iowa Agricultural Extension Service.

Beal, George M., and Everett M. Rogers. 1960. *The Adoption of Two Farm Practices in a Central Iowa Community.* Special Report 26. Ames: Iowa Agricultural Experiment Station.

Beal, George M., Everett M. Rogers, and Joe M. Bohlen. 1957. "Validity of the Concept of Stages in the Adoption Process." *Rural Sociology* 22 (2): 166–68.

Beck, Ulrich, Anthony Giddens, and Scott Lash. 1994. *Reflexive Modernization.* Cambridge: Polity Press.

Becker, Elizabeth. 2002. "Big Farms Making a Mess of U.S. Waters, Cities Say." *New York Times,* Feb. 10. Accessed Jan. 13, at http://www.nytimes.com/2002/02/10/politics/10FARM.html?ex=1014317949&ei=1& en=3ef82ae2b8c1621d.

Beeman, Randal S., and James A. Pritchard. 2001. *A Green and Permanent Land: Ecology and Agriculture in the Twentieth Century.* Lawrence: University Press of Kansas.

Bell, Michael M. 1992. "The Fruit of Difference: The Rural-Urban Continuum as a System of Identity." *Rural Sociology* 57 (1): 65–82.

———. 1994. *Childerley: Nature and Morality in a Country Village.* Chicago: University of Chicago Press.

———. 1998a. *An Invitation to Environmental Sociology.* Thousand Oaks, Calif.: Pine Forge Press.

———. 1998b. "Culture as Dialog." In *Bakhtin and the Human Sciences: No Last Words,* ed. Michael M. Bell and Michael Gardiner, 49–62. London: Sage.

———. 1998c. "The Dialogue of Solidarities, or Why the Lion Spared Androcles." *Sociological Focus* 31 (2): 181–99.

———. 1999. "The Social Construction of Farm Crises." Paper presented at the annual meeting of the Rural Sociological Society, Chicago.

———. 2002. "Sentences and Commitments." *International Journal of Humanities and Peace* 18 (1): 58.

———. 2003. "Dialogue and Isodemocracy: Creating the Social Conditions of Good Talk." In *Walking Toward Justice: Democratization in Rural Life,* ed. Michael M. Bell and Frederick Hendricks, with Azril Bacal. Amsterdam: JAI/Elsevier.

———. 2004. "Farms." In *Patterned Ground: Ecologies and Geographies of Nature and Culture,* ed. Stephan Harrison, Steve Pile, and Nigel Thrift, 142–43. London: Reaktion Books.

Bell, Michael M., and Iverson Griffin Jr. 2002. "Heritas: The Construction and Reconstruction of Communities of Descent." Paper presented at the annual meeting of the Association of American Geographers, Los Angeles.

Bentley, Jeffery W., G. Rodriguez, and A. Gonzalez. 1994. "Science and People: Honduran Campesinos and Natural Pest Control Inventions." *Agriculture and Human Values* 11 (2–3): 178–82.

Berry, Wendell. 1977. *The Unsettling of America: Culture and Agriculture.* San Francisco: Sierra Club.

———. 1990. "The Pleasures of Eating." In Wendell Berry, *What Are People For?* San Francisco: North Point Press.

Bird, Elizabeth Ann, Gordon Bultena, and John C. Gardner. 1995. *Planting the Future: Developing an Agriculture that Sustains Land and Community.* Ames: Iowa State University Press.

Blank, Steven C. 1998. *The End of Agriculture in the American Portfolio.* Westport, Conn.: Quorum Books.

Blisard, Noel. 2001. *Food Spending in American Households, 1997–98.* USDA Statistical Bulletin No. 972. Electronic Report from the Economic Research Service. Accessed Nov. 12, 2003, at http://www.ers.usda.gov/publications/sb972/.

Bogdan, Robert, and Steven J. Taylor. 1990. "Looking at the Bright Side: A Positive Approach to Qualitative Policy and Evaluation Research." *Qualitative Sociology* 13 (2): 183–92.

Bohlen, Joe M. 1967. "Needed Research on Adoption Models." *Sociologia Ruralis* 7 (2): 113–29.

Bonanno, Alessandro, ed. 1994. *From Columbus to ConAgra: The Globalization of Agriculture and Food.* Lawrence: University Press of Kansas.

Bormann, F. Herbert, Diana Balmori, and Gordon T. Geballe. 1993. *Redesigning the American Lawn: A Search for Environmental Harmony.* New Haven: Yale University Press.

Boserup, Ester. [1970] 1989. *Woman's Role in Economic Development.* London: Earthscan.

Bourdieu, Pierre, and Loic Wacquant. 1992. *An Invitation to Reflexive Sociology.* Chicago: University of Chicago Press.

Bové, José, and Francois Dufour. 2001. *The World Is Not for Sale: Farmers Against Junk Food.* Trans. Anna de Casparis. London: Verso.

Brandth, B. 1995. "Rural Masculinity in Transition: Gender Images in Tractor Advertisements." *Journal of Rural Studies* 11: 123–33.

Brecher, Jeremy. 1994. *Global Village or Global Pillage: Economic Reconstruction from the Bottom Up.* Boston: South End Press.

Brewster, Clarence, James Kleibenstein, Mary Honeyman, and Arlie Penner. N.d. "Cost of Finishing Pigs in Hoop and Confinement Facilities." Iowa State University Report ASL-R1686. Accessed June 27, 2002, at http://www.extension.iastate.edu/ipic/reports/99swinereports/asl-1686.pdf.

Brookfield, Harold. 2001. *Exploring Agrodiversity.* New York: Columbia University Press.

Bryson, Bill. 1989. *The Lost Continent.* HarperCollins.

Bureau of Labor Statistics. N.d. "Industry at a Glance: Construction." Accessed May 23, 2003, at http://stats.bls.gov/iag/iag.construction.htm.

———. 2001. *The Employment Situation: January, 2001.* Washington, D.C.: U.S. Department of Labor. Accessed June 27, 2002, at http://www.bls.census.gov/cps/pub/empsit_jan2001.htm.

Busch, L., and W. B. Lacy. 1983. *Science, Agriculture, and the Politics of Research.* Boulder: Westview Press.

———, eds. 1986. *The Agricultural Scientific Enterprise: A System in Transition.* Boulder: Westview Press.

Buttel, Frederick H., and L. Busch. 1988. "The Public Agricultural Research System at the Crossroads." *Agricultural History* 62 (2): 303–24.

Buttel, Frederick H., Olaf F. Larson, and Gilbert W. Gillespie Jr. 1990. *The Sociology of Agriculture.* New York: Greenwood Press.

Campbell, H. 2000. "The Glass Phallus: Pub(lic) Masculinity and Drinking in Rural New Zealand." *Rural Sociology* 65: 562–81.

Campbell, H., and Michael M. Bell. 2000. "The Question of Rural Masculinities." *Rural Sociology* 65: 532–46.

Carolan, Michael, and Michael M. Bell. 2003. "In Truth We Trust: Discourse, Phenomenology, and the Social Relations of Knowledge in an Environmental Dispute." *Environmental Values* 12 (2): 225–45.

Chambers, Robert. 1994. "The Origins and Practice of Participatory Rural Appraisal." *World Development* 22 (7): 953–69.

Chavas, Jean-Paul, and Michael Aliber. 1993. "An Analysis of Economic Efficiency in Agriculture: A Nonparametric Approach." *Journal of Agricultural and Resource Economics* 18 (1): 1–16.

Cocchi, Horacio, Boris E. Bravo-Ureta, and Stephen Cooke. 1998. "A Growth Accounting Analysis of Cost Efficiency in Milk Production for Six Northern States in the United States." *Canadian Journal of Agricultural Economics* 46 (3): 287–96.

Cochrane, Willard W. 1958. *Farm Prices: Myth and Reality*. Minneapolis: University of Minnesota Press.

———. 1965. *The City Man's Guide to the Farm Problem*. Minneapolis: University of Minnesota Press.

Collins, Patricia Hill. 1986. "Learning from the Outsider Within: The Sociological Significance of Black Feminist Thought." *Social Problems* 33 (6): s14–s32.

Collinson, M. 2000. *A History of Farming Systems Research*. Wallingford, UK: CABI Publishing.

Comstock, Gary, ed. 1986. *Is There a Moral Obligation to Save the Family Farm?* Ames: Iowa State University Press.

Connell, R. W. 1995. *Masculinities*. Berkeley and Los Angeles: University of California Press.

Davidson, Osha Gray. [1990] 1996. *Broken Heartland: The Rise of America's Rural Ghetto*. Exp. ed. Iowa City: University of Iowa Press.

De Beauvoir, Simone. [1949] 1989. *The Second Sex*. Trans. H. M. Parshley. New York: Vintage Books.

Degregori, Thomas R. 2001. *Agriculture and Modern Technology: A Defense*. Ames: Iowa State University Press.

Delgado, Richard, and Jean Stefancic. 1998. "Critical Race Theory: Past, Present, and Future." In *Legal Theory at the End of the Millennium*, ed. M. D. A. Freeman, 467–91. Oxford: Oxford University Press.

Depew, David, and Robert Hollinger. [1995] 1999. *Pragmatism: From Progressivism to Postmodernism*. Westport, Conn.: Praeger.

Dewey, John. 1928. *The Philosophy of John Dewey*, ed. Joseph Ratner. New York: Henry Holt and Co.

———. 1939. "Experience, Knowledge, and Value: A Rejoinder." In *The Philosophy of John Dewey*, ed. Paul Arthur Schlipp, 517–608. Evanston: Northwestern University Press.

———. 1988. *The Later Works, 1925–1953*. 17 vols. Ed. Jo Ann Boydston. Carbondale: Southern Illinois University Press.

Dudley, Kathryn Marie. 2000. *Debt and Dispossession: Farm Loss in America's Heartland*. Chicago: University of Chicago Press.

Duffy, Michael. 1999. "Farmers' Thoughts on Sustainable Agriculture." *Leopold Letter* 11 (1). Accessed June 6, 2003, at http://www.ag.iastate.edu/centers/leopold/newsletter/99-1leoletter/99-1lcsurvey.html.

———. 2001. "Who Benefits from Biotechnology?" Paper presented at the American Seed Trade Association meeting, Dec. 5–7, Chicago.

Duffy, Michael, and Matt Ernst. 1999. "Does Planting GMO Seed Boost Farmers' Profits?" *Leopold Letter* 11 (3), Leopold Center for Sustainable Agriculture. Accessed June 27, 2002, at http://www.ag.iastate.edu/centers/leopold/newsletter/99-3leoletter/99-3gmoduffy.html.

Duffy, Michael, and Darnell Smith. 2002. *Estimated Costs of Crop Production in Iowa—2002*. Accessed May 25, 2002, at http://www.econ.iastate.edu/faculty/duffy/.

Duneier, Mitchell. 1999. *Sidewalk*. New York: Farrar, Straus, and Giroux.

Duneier, Mitchell, and Harvey Molotch. 1999. "Talking City Trouble: Interactional Vandalism, Social Inequality, and the Urban Interaction Problem." *American Journal of Sociology* 104 (5): 263–95.

Durning, Alan T. 1992. *How Much is Enough? Consumer Society and the Future of the Earth*. New York: W. W. Norton.

Dupuis, Melanie. 2000. "Not in My Body: rBGH and the Rise of Organic Milk." *Agriculture and Human Values* 17: 285–95.

Economic Research Service. 2003. *Food CPI, Prices, and Expenditures*. Accessed June 21, 2003, at http://www.ers.usda.gov/Briefing/CPIFoodAndExpenditures/Data/table7.htm.

Eisenberg, Deborah Thompson. N.d. "The Feudal Lord in the Kingdom of Big Chicken: Contracting and Worker Exploitation in the Poultry Industry." Public Justice Center. Accessed May 19, 2002, at http://www.nelp.org/appendices/swi/a/eisenberg.pdf.

Environmental Working Group. 2000. *Freedom to Farm in Iowa, 1996–1998*. Washington, D.C.: Environmental Working Group. Accessed June 27, 2002, at http://www.ewg.org/reports/freetofarm/IAfreedom.html.

———. 2002. *EWG Farm Subsidy Database*. Accessed May 25, 2002, at http://gsi.ewg.org/farmbill.acgi$farmbill2?regtype=state&state=IA&submitte d=true&whichForm=reg.

Epstein, Cynthia Fuchs. 1988. *Deceptive Distinctions*. New Haven: Yale University Press.

Exner, Rick, and Richard Thompson. 1998. *The Paired-Comparison: A Good Design for Farmer-Managed Trials*. Accessed June 16, 2002, at http://www.pfi.iastate.edu/OFR/OFR_worksheet.htm.

Feldman, Shelley, and Rick Welsh. 1995. "Feminist Knowledge Claims, Local Knowledge, and Gender Divisions of Agricultural Labor: Constructing a Successor Science." *Rural Sociology* 60 (1): 23–43.

Fenton, Natalie. 2000. "The Problematics of Postmodernism for Feminist Media Studies." *Media Culture & Society* 22 (6): 723–741.

Fliegel, Frederick C. 1993. *Diffusion Research in Rural Sociology: The Record and Prospects for the Future*. Westport, Conn.: Greenwood Press.

Flora, Cornelia Butler. 1985. "Women and Agriculture." *Agriculture and Human Values* 2: 5–12.

———. 1992. "Reconstructing Agriculture: The Case for Local Knowledge." *Rural Sociology* 57: 92–97.

Food Research and Action Center. 2003. *Hunger in the U.S.* Accessed June 20, 2003 at http://www.frac.org/html/hunger_in_the_us/hunger_index.html.

Foucault, Michel. [1975] 1979. *Discipline and Punish: The Birth of the Prison*. New York: Vintage Books.

———. 1980. *Power/Knowledge: Selected Interviews and Other Writings, 1972–1977*. Trans. Colin Gordon. New York: Random House.

———. 1984. "What Is an Author?" In *The Foucault Reader,* ed. Paul Rabinow, 101–20. New York: Pantheon Books.

———. 1985. *The Use of Pleasure.* Vol. 2 of *The History of Sexuality.* Harmondsworth, UK: Penguin Books.

———. 1986. *The Care of the* Self. Vol. 3 of *The History of Sexuality.* New York: Pantheon Books.

———. 2000. *Power.* Vol. 3 of *Essential Works of Foucault, 1954–1984.* Trans. Robert Hurley et al. New York: New Press.

Francis, Charles A. 1990. "Sustainable Agriculture: Myths and Realities." *Journal of Sustainable Agriculture* 1 (1): 97–106.

Friedland, Williams H. 2002. "Agriculture and Rurality: Beginning the Final Separation?" *Rural Sociology* 67 (3): 350–371.

Friedmann, Harriet. 2003. "Eating in the Gardens of Gaia: Envisioning Polycultural Communities." In *Fighting for the Farm: Rural America Transformed,* ed. Jane Adams, 252–73. Philadelphia: University of Pennsylvania Press.

Freire, Paulo. [1970] 1999. *Pedagogy of the Oppressed.* New York: Continuum.

Funes, Fernando, Luis Garcia, Martin Bourque, Nilda Perez, and Peter Rosset. 2002. *Sustainable Agriculture and Resistance: Transforming Food Production in Cuba.* Oakland, Calif.: Food First Books.

Gardiner, Michael E. 2000. *Critiques of Everyday Life.* London: Routledge.

Gardiner, Michael E., and Michael M. Bell. 1998. "Bakhtin and the Human Sciences: An Introduction." In *Bakhtin and the Human Sciences: No Last Words,* ed. Michael M. Bell and Michael Gardiner, 1–12. London: Sage.

Gaventa, John. 1993. "The Powerful, the Powerless, and the Experts: Knowledge Struggles in an Information Age." In *Voices of Change: Participatory Research in the United States and Canada,* ed. Peter Park, Mary Brydon-Miller, Budd Hall, and Ted Jackson, 21–40. London: Bergin and Garvey.

Geertz, Clifford. 1973. "Thick Description: Toward an Interpretive Theory of Culture." In Clifford Geertz, *The Interpretation of Cultures,* 3–20. New York: Basic Books.

Giddens, Anthony. N.d. "Frequently Asked Questions (FAQs)." Accessed June 6, 2003, at http://www.lse.ac.uk/Giddens/FAQs.htm.

———. 1984. *The Constitution of Society: Outline of the Theory of Structuration.* Cambridge, UK: Polity Press.

Gilbert, Jess. 2003. "Low Modernism and the Agrarian New Deal: A Different Kind of State." In *Fighting for the Farm: Rural America Transformed,* ed. Jane Adams, 129–46. Philadelphia: University of Pennsylvania Press.

Gieryn, Thomas F. 2000. "A Space for Place in Sociology." *Annual Review of Sociology* 26: 463–96.

Gilligan, Carol. [1982] 1993. *In a Different Voice: Psychological Theory and Women's Development.* 2d ed. Cambridge: Harvard University Press.

Gliessman, Stephen R. 1998. *Agroecology: Ecological Processes in Sustainable Agriculture.* Chelsea, Mich.: Ann Arbor Press.

Goffman, Erving. [1974] 1986. *Frame Analysis: An Essay on the Organization of Experience.* Boston: Northeastern University Press.

Gordon, Colin. 1980. "Afterword." In Michel Foucault, *Power/Knowledge: Selected Interviews and Other Writings, 1972–1977,* trans. Colin Gordon, 229–59. New York: Random House.

Goudy, Willis, Sandra Charvat Burke, Seung-pyo Hong, Liu Dong Wang, Liu Qiang, and John Wallize. 1996. *Rural/Urban Transitions in Iowa.* Iowa State University, Department of Sociology, Census Services, Report CS96–4.

Gray, Osha. [1990] 1996. *Broken Heartland: The Rise of America's Rural Ghetto.* Iowa City: University of Iowa Press.

Haraway, Donna Jeanne. 1991. *Simians, Cyborgs, and Women: The Reinvention of Nature.* New York: Routledge.

Harper, Douglas. 2001. *Changing Works: Visions of a Lost Agriculture.* Chicago: University of Chicago Press.

Hart, John Fraser. 1991. *The Land That Feeds Us.* New York: W. W. Norton.

Hassanein, Neva. 1999. *Changing the Way America Farms: Knowledge and Community in the Sustainable Agriculture Movement.* Lincoln: University of Nebraska Press.

Hassanein, Neva, and Jack Kloppenburg Jr. 1995. "Where the Grass Grows Again: Knowledge Exchange in the Sustainable Agriculture Movement." *Rural Sociology* 60 (4): 721–40.

Hauser, N., and C. Keleusal, eds. 1992. *The Essential Peirce.* Bloomington: Indiana University Press.

Heffernan, William, with Mary Hendrickson and Robert Gronski. 1999. *Consolidation in the Food and Agriculture System: Report to the National Farmers' Union.* Accessed Jan. 13, 2003, at http://foodcircles.missouri.edu/pub.htm.

Hendrickson, Mary, and William Heffernan. 2002. *Concentration of Agricultural Markets: February, 2002.* Accessed Jan. 14, 2002, at http://www.nfu.org/index.cfm.

Hightower, Jim. 1973. *Hard Tomatoes, Hard Times: A Report of the Agribusiness Accountability Project on the Failure of America's Land Grant College Complex.* Cambridge, Mass.: Schenkman.

Hightower, Jim, and Susan DeMarco. [1976] 1986. "Hard Tomatoes, Hard Times: The Failure of the Land Grant College Complex." In *Is There a Moral Obligation to Save the Family Farm?* ed. Gary Comstock, 135–52. Ames: Iowa State University Press.

Hildebrand, David L. 2003. *Beyond Realism and Antirealism: John Dewey and the Neopragmatists.* Nashville: Vanderbilt University Press.

Hochschild, Arlie. 1997. *The Time Bind: When Work Becomes Home and Home Becomes Work.* New York: Metropolitan Books.

Hochschild, Arlie, with Annie Machung. [1989] 1997. *The Second Shift.* New York: Avon Books.

Iowa Office of the National Agricultural Statistics Service. 2003. *Number of Farms, Land in Farms, and Average Farm Size, Iowa by County, 2001–2002.* Accessed June 20, 2003, at http://www.nass.usda.gov/ia/coest/farms01_02.txt.

Iowa Workforce Development. 2003. *Iowa Employment Situation: May 2003*. Accessed June 20, 2003, at http://www.iowaworkforce.org/news/XcNewsPlus.asp?cmd=view&articleid=81.

Jackson, Wes. 1985. *New Roots for Agriculture*. 2d ed. Lincoln: University of Nebraska Press.

Jackson, Dana. L., and Laura. L. Jackson, eds. 2002. *The Farm as Natural Habitat: Reconnecting Food Systems with Ecosystems*. Covelo, Calif.: Island Press.

Jackson-Smith, Douglas B. 1999. "Understanding the Microdynamics of Farm Structural Change: Entry, Exit, and Restructuring Among Wisconsin Family Farmers in the 1980s." *Rural Sociology* 64 (1): 66–91.

Jacobs, Jane. 2000. *The Nature of Economies*. New York: Modern Library.

Jaforullah, Mohammad, and John Whiteman. 1999. "Scale Efficiency in the New Zealand Dairy Industry: A Nonparametric Approach." *Australian Journal of Agricultural and Resource Economics* 43 (4): 523–41.

James, William. [1896] 1927. *The Will to Believe, and Other Essays in Popular Philosophy*. New York: Longmans, Green and Co.

———. [1925] 1953. *The Philosophy of William James*. New York: Modern Library.

Johnson, H. Thomas. 1992. *Relevance Regained: From Top-Down Control to Bottom-Up Empowerment*. New York: Free Press.

Jordan, Lowell S., et al. 1990. *Alternative Agriculture: Scientists' Review*. Ames: Council for Agricultural Science and Technology.

Kahn, Barbara E., and Leigh McAlister. 1997. *Grocery Revolution: The New Focus on the Consumer*. New York: Addison-Wesley.

Kaufman, Phillip R., James M. MacDonald, Steve M. Lutz, and David M. Smallwood. 1997. *Do the Poor Pay More for Food?* Economic Research Service, U.S. Department of Agriculture, Agricultural Economic Report 759.

Kemmis, Stephen, and Robin McTaggart. 2000. "Participatory Action Research." In *Handbook of Qualitative Research*, 2d ed., ed. Norman K. Denzin and Yvonna S. Lincoln, 567–606. Thousand Oaks, Calif.: Sage.

Kimmel, M., and A. Ferber. 2000. "'White Men Are This Nation': Right-Wing Militias and the Restoration of Rural American Masculinity." *Rural Sociology* 65: 582–604.

Kloppenburg, Jack Jr. 1991. "Social Theory and the De/Reconstruction of Agricultural Science: Local Knowledge for an Alternative Agriculture." *Rural Sociology* 56: 519–48.

Kloppenburg, Jack Jr., John Hendrickson, and George W. Stevenson. 1996. "Coming into the Foodshed." *Agriculture and Human Values* 13: 33–42.

Korten, David. 1999. *The Post-Corporate World: Life After Capitalism*. San Francisco: Berrett-Koehler Publishers.

Lasley, Paul, F. Larry Leistritz, Linda M. Lobao, and Katherine Meyer. 1995. *Beyond the Amber Waves of Grain: An Examination of Social and Economic Restructuring in the Heartland*. Boulder: Westview Press.

Law, R., H. Campbell, and J. Dolan, eds. 1999. *Masculinities in Aotearoa/New Zealand*. Palmerston North, N.Z.: Dunmore Press.

Lay, D. C. Jr., M. F. Haussmann, and M. J. Daniels. N.d. "A Comparison of the Behavior and Physiology of Pigs Raised in Hoop Structures Compared to a Non-Bedded Confinement System." Iowa State University, Rhodes Research and Development Farm Report. Accessed June 27, 2002, at http://www.ae.iastate.edu/hoop_structures/hoopfeld99.pdf.

Leipins, R. 2000. "Making Men: The Construction and Representation of Agriculture-Based Masculinities in Australia and New Zealand." *Rural Sociology* 65: 605–20.

Leopold, Aldo. [1949] 1966. *A Sand County Almanac, with Essays on Conservation from Round River.* New York: Sierra Club/Ballantine.

Liebman, Matt. 2001. "Weed Management: A Need for Ecological Approaches." In *Ecological Management of Agricultural Weeds,* ed. M. Liebman, C. L. Mohler, and C. P. Staver, 1–39. Cambridge: Cambridge University Press.

Little, J., and O. Jones. 2000. "Masculinity, Gender, and Rural Policy." *Rural Sociology* 65: 621–39.

Lobao, Linda. 1990. *Locality and Inequality: Farm and Industry Structure and Socioeconomic Conditions.* Albany: State University of New York Press.

Logsdon, Gene. 1994. *At Nature's Pace: Farming and the American Dream.* New York: Pantheon.

Lorber, Judith. 1994. *Paradoxes of Gender.* New Haven: Yale University Press.

Lyson, Thomas A. 2000. "Moving Toward Civic Agriculture." *Choices* 3: 42–45.

Malone, Bill. 2002. *Don't Get Above Your Raisin': Country Music and the Southern Working Class.* Urbana: University of Illinois Press.

Marx, Karl. [1852] 1972. "The Eighteenth Brumaire of Louis Bonaparte." In *The Marx-Engels Reader,* ed. Robert C. Tucker, 436–525. New York: W. W. Norton.

McEvoy, Arthur F. [1986] 1990. *The Fisherman's Problem: Ecology and Law in the California Fisheries, 1850–1980.* New York: Cambridge University Press.

McMichael, Philip. 2000. "The Power of Food." *Agriculture and Human Values* 17: 21–33.

Meares, Alison. 1997. "Making the Transition from Conventional to Sustainable Agriculture: Gender, Social Movement Participation, and Quality of Life on the Family Farm." *Rural Sociology* 62 (1): 21–47.

Mittelberg, David, and Mary C. Waters. 1992. "The Process of Ethnogenesis Among Haitian and Isreali Immigrants in the United States." *Ethnic and Racial Studies* 15 (3): 412–35.

Mooney, Patrick H. 1988. *My Own Boss? Class, Rationality, and the Family Farm.* Boulder: Westview Press.

Morson, Gary Saul, and Caryl Emerson. 1990. *Mikhail Bakhtin: Creation of a Prosaics.* Stanford: Stanford University Press.

Musambira, George W. 2000. "A Comparison of Modernist and Postmodernist Accounts of Cross-Cultural Communication Between African Societies and the United States." *Howard Journal of Communications* 11 (2): 145–161.

Myrdal, Gunnar. 1944. *An American Dilemma: The Negro Problem and Modern Democracy.* London: Harper and Bros.

National Agricultural Statistics Service. 1999. *1997 Census of Agriculture, Iowa.* Washington, D.C.: U.S. Government Printing Office.

———. 2000. *Agricultural Statistics, 2000.* Washington, D.C.: U.S. Government Printing Office.

———. 2002. *Agricultural Statistics, 2002.* Washington, D.C.: U.S. Government Printing Office.

National Research Council. 1989. *Alternative Agriculture.* Washington, D.C.: National Academy Press.

Nippert-Eng, Christena E. 1996. *Home and Work: Negotiating Boundaries Through Everyday Life.* Chicago: University of Chicago Press.

Noller, P. 2000. "Globalization, Space, and Society: Elements of a Modern Sociology of Space." *Berliner Journal für Soziologie* 10 (1): 21–48.

Nord, Mark, Margaret Andrews, and Steven Carlson. 2003. *Household Food Security in the United States, 2002.* U.S. Department of Agriculture, Food and Rural Economics Division, Economic Research Service. Food Assistance and Nutrition Research Report No. 35.

O'Connor, James. 1973. *The Fiscal Crisis of the State.* New York: St. Martin's Press.

———. 1984. *Accumulation Crisis.* Oxford: Basil Blackwell.

Omi, Michael, and Howard Winant. [1986] 1994. *Racial Formation in the United States.* 2d ed. New York: Routledge.

Organic Trade Association. 2003. "Industry Statistics and Projected Growth." Accessed Oct. 24, 2003 at http://www.ota.com/organic/mt/business.html.

Otto, Daniel, David Swenson, and Mark Immerman. 1997. *The Role of Agriculture in the Iowa Economy.* Iowa State University, Department of Economics report. Accessed April 26, 2000, at http://www.econ.iastate.edu/outreach/agriculture/AgImpactStudy/reports/state.htm.

Otto, Daniel, and John Lawrence. 2002. *The Iowa Pork Industry 2000: Trends and Economic Performance.* Iowa State University, Department of Economics. Accessed May 21, 2002, at http://www.econ.iastate.edu/faculty/Lawrence/Acrobat/iowa%20hogs2002.pdf.

Padel, Susanne. 2001. "Conversion to Organic Farming: A Typical Example of the Diffusion of an Innovation?" *Sociologia Ruralis* 41 (1): 40–61.

Peirce, Charles S. 1878. "How to Make Our Ideas Clear." *Popular Science Monthly* 12: 286–302.

Penning-Rowsell, Edmund C., and David Lowenthal. 1986. *Landscape Meaning and Values.* London: Allen and Unwin.

Peter, G., M. Bell, S. Jarnagin, and D. Bauer. 2000. "Coming Back Across the Fence: Masculinity and the Transition to Sustainable Agriculture." *Rural Sociology* 65: 215–33.

Pile, Stephen. 1990. *The Private Farmer: Transformation and Legitimation in Advanced Capitalist Agriculture.* Aldershot, England: Dartmouth Publishing Co.

Pollack, Andrew. 2003. "Widely Used Crop Herbicide Is Losing Weed Resistance." *New York Times,* Jan. 14. Accessed Jan. 16, 2003, at http://www.nytimes.com/2003/01/14/business/14WEED.html.

Pollan, Michael. 2001. "Behind the Industrial-Organic Complex." *New York Times Magazine*, May 13. Accessed June 19, 2003, at http://www.nytimes.com/2001/05/13/magazine/13ORGANIC.html.

Press, Andrea L. 2000. "Recent Developments in Feminist Communication Theory: Difference, Public Sphere, Body and Technology." In *Mass Media and Society*, ed. James Curran and Michael Gurevitch, 27–43. London: Arnold.

Putnam, Robert D. 2000. *Bowling Alone: The Collapse and Revival of American Community*. New York: Simon and Schuster.

Rawls, John. 1972. *A Theory of Justice*. Cambridge: Harvard University Press.

Richards, Timothy, and Jeffrey R. Scott. 2000. "Efficiency and Economic Performance: An Application of the MIMIC Model." *Journal of Agricultural and Resource Economics* 25 (1): 232–51.

Rogers, Everett M. 1995. *Diffusion of Innovations*. 4th ed. New York: Free Press.

Roling, Niels G., and M. A. E. Wagenmakers, eds. 1998. *Facilitating Sustainable Agriculture: Participatory Learning and Adaptive Management in Times of Environmental Uncertainty*. Cambridge: Cambridge University Press.

Roling, Niels G., and Janice Jiggins. 1998. "The Ecological Knowledge System." In *Facilitating Sustainable Agriculture: Participatory Learning and Adaptive Management in Times of Environmental Uncertainty*, ed. N. G. Roling and M. A. E. Wagenmakers, 283–311. Cambridge: Cambridge University Press.

Rorty, Richard. 1999. *Philosophy and Social Hope*. London: Penguin Books.

Ryan, Bryce, and Neal C. Gross. 1943. "The Diffusion of Hybrid Seed Corn in Two Iowa Communities." *Rural Sociology* 8: 15–24.

Sachs, Carolyn. 1996. *Gendered Fields: Rural Women, Agriculture, and Environment*. Boulder: Westview Press.

Salamon, Sonya. 1985. "Ethnic Communities and the Structure of Agriculture." *Rural Sociology* 50: 323–40.

———. 1992. *Prairie Patrimony: Family, Farming, and Community in the Midwest*. Chapel Hill: University of North Carolina Press.

———. 2002. *Newcomers, Old Towns: Community Change in the Postagrarian Midwest*. Chicago: University of Chicago Press.

Sands, James K., and Howard R. Holden. 1998. *Iowa Agricultural Statistics, 1998*. Des Moines: U.S. Department of Agriculture and Iowa Farm Bureau.

———. 2000. *Iowa Agricultural Statistics, 2000*. Des Moines: U.S. Department of Agriculture and Iowa Farm Bureau.

Schegloff, Emanuel A. 1996. "Confirming Allusions: Toward an Empirical Account of Action." *American Journal of Sociology* 102 (1): 161–216.

Schlosser, Eric. 2002. *Fast Food Nation: The Dark Side of the All-American Meal*. New York: Perennial.

Schnaiberg, Alan. 1980. *The Environment, from Surplus to Scarcity*. New York: Oxford University Press.

Schor, Juliet. 1999. *The Overspent American: Upscaling, Downshifting, and the New Consumer*. New York: HarperCollins.

Scott, James C. 1985. *Weapons of the Weak: Everyday Forms of Peasant Resistance.* New Haven: Yale University Press.

———. 1990. *Domination and the Arts of Resistance: Hidden Transcripts.* New Haven: Yale University Press.

Sedgwick, Eve Kosofsky. 1990. *Epistemology of the Closet.* Berkeley and Los Angeles: University of California Press.

Sen, Amartya. 1992. *Inequality Reexamined.* Cambridge: Harvard University Press.

Simmel, Georg. [1908] 1971. "The Stranger." In *On Individuality and Social Forms: Selected Writings,* ed. Donald N. Levine, 143–49. Chicago: University of Chicago Press.

Smith, Dorothy E. 1987. *The Everyday World as Problematic: A Feminist Sociology.* Boston: Northeastern University Press.

Stein, Arlene. 2001. *The Stranger Next Door: The Story of a Small Community's Battle over Sex, Faith, and Civil Rights.* Boston: Beacon Press.

Stoecker, Randy, and Edna Bonacich. 1992. "Why Participatory Research? Guest Editor's Introduction." *American Sociologist* 23 (4): 5–14.

Stolen, K. A. 1995. "The Gentle Exercise of Male Power in Rural Argentina." *Identities: Global Studies in Culture and Power* 2: 385–406.

Strange, Marty. 1988. *Family Farming: A New Economic Vision.* Lincoln: University of Nebraska Press.

Strozier, Robert M. 2002. *Foucault, Subjectivity, and Identity: Historical Constructions of Subject and Self.* Detroit: Wayne State University Press.

Teyssot, Georges. 1999. *The American Lawn.* New York: Princeton Architectural Press.

Thompson, Dick, Sharon Thompson, and Rex Thompson. 2001. *Alternatives in Agriculture: 2001 Report.* Boone, Iowa: Thompson On-Farm Research.

———. 2002. *Alternatives in Agriculture: 2002 Report.* Boone, Iowa: Thompson On-Farm Research.

Thu, Kendall M., and E. Paul Durrenberger. 1998. *Pigs, Profits, and Rural Communities.* Albany: State University of New York Press.

Tickamyer, Ann, and Janet Bokemeier. 1988. "Sex Differences in Labor Market Experiences." *Rural Sociology* 53 (2): 166–89.

Tweeten, Luther G. 2003. *Terrorism, Radicalism, and Populism in Agriculture.* Ames: Iowa State University Press.

United Nations Population Fund. 1999. *State of the World's Population, 1999.* Accessed June 18, 2003, at http://www.unfpa.org/swp/1999/chapter2d.htm.

U.S. Census Bureau. 1999. *Money Income in the United States: 1999.* Current Population Reports P60–209. Washington, D.C.: U.S. Government Printing Office.

U.S. Department of Agriculture. 1997. *1997 Census of Agriculture: State Profile: Iowa.* Accessed June 15, 2002, at http://www.nass.usda.gov/census/census97/profiles/ia/iapbst.pdf.

———. 1998a. *1997 Census of Agriculture: Ranking of States and Counties.* Washington, D.C.: U.S. Government Printing Office.

———. 1998b. *Agriculture Fact Book, 1998.* Washington, D.C.: U.S. Government Printing Office.

———. 2002. *Farmers Market Facts!* Accessed June 19, 2003, at http://www.ams.usda.gov/farmersmarkets/facts.htm.

USDA Economic Research Service. 2003. "Food Marketing and Price Spreads: USDA Marketing Bill." Accessed Nov. 12, 2003, at http://www.ers.usda.gov/briefing/foodpricespreads/bill/.

Valdivia, C., and J. Gilles. 2001. "Gender and Resource Management: Households and Groups, Strategies and Transitions." *Agriculture and Human Values* 18 (1): 5–9.

Valente, Thomas W., and Everett M. Rogers. 1995. "The Origins and Development of the Diffusion of Innovations Paradigm as an Example of Scientific Growth." *Science Communication* 16 (3): 242–73.

Van den Berghe, Pierre L. [1967] 1978. *Race and Racism: A Comparative Perspective.* New York: John Wiley.

Voisin, Andre. [1959] 1988. *Grass Productivity.* Washington, D.C.: Island Press.

Volosinov, V. N. [1929] 1973. *Marxism and the Philosophy of Language.* Trans. Ladislav Matejka and I. R. Titunik. New York: Seminar Press.

Walzer, M. 1983. *Spheres of Justice: A Defense of Pluralism and Equality.* New York: Basic Books.

Wardell, Mark. 1992. "Changing Organizational Forms: From the Bottom Up." In *Rethinking Organization: New Directions in Organization Theory and Analysis,* ed. Michael Reed and Michael Hughes. London: Sage.

Webber, Melvin. 1963. "Order in Diversity: Community Without Propinquity." In *Cities and Space,* ed. Lowdon Wingo, 25–54. Baltimore: Johns Hopkins University Press.

Wells, Betty, and Shelly Gradwell. 2001. "Gender and Resource Management: Community Supported Agriculture as Caring Practice." *Agriculture and Human Values* 18 (1): 107–19.

West, Candace, and Don Zimmerman. 1987. "Doing Gender." *Gender and Society* 1: 125–51.

Whatmore, Sarah. 1991. *Farming Women: Gender, Work, and Family Enterprise.* London: Macmillan.

Whyte, William Foote. 1985. "Review of *Technology and Social Change in Rural Areas.*" *Contemporary Sociology* 14: 65.

Williams, Raymond. 1973. *The Country and the City.* London: Chatto and Windus.

Wolf, Steven, and David Zilberman, eds. 2001. *Knowledge Generation and Technical Change: Institutional Innovation in Agriculture.* Boston: Kluwer.

Woodward, R. 2000. "Warrior Heroes and Little Green Men: Soldiers, Military Training, and the Construction of Rural Masculinities." *Rural Sociology* 65: 640–57.

World Almanac. 1997. *World Almanac and Book of Facts, 1997.* New York: World Almanac Books.

Zinkand, Dan. 2001. "Of Pigs and Prairies." *Iowa Alumnae Magazine.* Accessed June 16, 2002, at http://www.iowalum.com/magazine/august01/pigs.html.

Zita, Jacquelyn N. 1998. *Body Talk: Philosophical Reflections on Sex and Gender.* New York: Columbia University Press.

INDEX

Page numbers appearing in *italic* type refer to photographs or figures.